PRACTICAL STATISTICS FOR BUSINESS

An Introduction to Business Statistics

Ruth Ravid
Perry Haan

University Press of America,® Inc.
Lanham · Boulder · New York · Toronto · Plymouth, UK

Copyright © 2008 by
University Press of America,® Inc.
4501 Forbes Boulevard
Suite 200
Lanham, Maryland 20706
UPA Acquisitions Department (301) 459-3366

Estover Road
Plymouth PL6 7PY
United Kingdom

Library of Congress Control Number: 2007933326
ISBN-13: 978-0-7618-3884-5 (paperback : alk. paper)
ISBN-10: 0-7618-3884-8 (paperback : alk. paper)

Table of Contents

Part Five: Inferential Statistics

Chapter 10: *t* Test 119

Hypotheses for *t* Tests 119
Using the *t* Test 121
t Test for Independent Samples 121
   An Example of a *t* Test for Independent Samples 122
t Test for Paired Samples 125
   An Example of a *t* Test for Paired Samples 126
t Test for a Single Sample 128
   An Example of a Single Sample *t* Test 128
SUMMARY 130

Chapter 11: Analysis of Variance 133

One-Way ANOVA 135
   Conceptualizing the One-Way ANOVA 135
   Hypotheses for a One-Way ANOVA 138
   The ANOVA Summary Table 138
   Further Interpretation of the *F* Ratio 139
   An Example of a One-Way ANOVA 140
   Post Hoc Comparisons 142
Two-Way ANOVA 143
   Conceptualizing the Two-Way ANOVA 143
   Hypotheses for the Two-Way ANOVA 144
   Graphing the Interaction 144
   The Two-Way ANOVA Summary Table 147
   An Example of a Two-Way ANOVA 147
SUMMARY 149

Chapter 12: Chi Square Test 153

Assumptions for the Chi Square Test 154
The Goodness of Fit Chi Square Test 154
   Equal Expected Frequencies 155
   Unequal Expected Frequencies 156
The Chi Square Test of Independence 157
SUMMARY 159

Part Six: Reliability and Validity

Chapter 13: Reliability 163

Understanding the Theory of Reliability 163
Methods of Assessing Reliability 164
   Test-Retest Reliability 164
   Alternate Forms Reliability 165
   Measures of Internal Consistency 165

Part Seven: Conducting Your Own Research

Preface

Practical Statistics for Business: An Introduction to Business Statistics, as the title indicates, was written specifically for business students. The book focuses on the application of research and statistics to business and a special effort was made to present examples relevant to those in all business disciplines. The book offers a clear and easy-to-follow text for students in a basic statistics course. The book can also serve as a resource for businesspeople and business educators who want to acquire the skills that are needed to conduct research and to study their own business settings.

There are seven parts in the book: *Introduction, Descriptive Statistics, The Normal Curve and Standard Scores, Measuring Relationships, Inferential Statistics, Reliability and Validity*, and *Conducting Your Own Research*. Each chapter ends with a summary.

A workbook to accompany the textbook was created (due out in 2008) for those who want to review, practice, and apply the statistical concepts and materials presented in the textbook. In order to provide students with immediate feedback and to allow them to confirm their answers to the exercises in the workbook, answers are found at the end of each chapter in the workbook. The workbook includes exercises to allow students to perform data analysis using the computer.

Ruth Ravid is the author of the highly successful *Practical Statistics for Educators* (2005; University Press of America), which is currently in its third edition. She wishes to dedicate the book to Dan, her late husband. Co-author **Perry Haan** wants to thank his wife, Meg Greenwald-Haan, for typesetting and proofing the text. Without her help, the text would not have been completed.

Part One

Introduction

Chapter 1

An Overview of Business Research

For many people, the term *research* conjures up a picture of a lab, researchers working at computers and "crunching numbers," or mice being injected with experimental drugs. Clearly, this is not what we mean by this term. In this book, we define **research** as a systematic inquiry that includes data collection and analysis. The goal of research is to describe, explain, or predict present or future phenomena. There are several ways to classify research into categories and each way looks at research from a different perspective. Research may be classified as: (a) *basic* (*pure*), *applied*, or *action* research; (b) *quantitative* or *qualitative* research; and (c) *experimental* or *nonexperimental* research.

Basic (Pure), Applied, and Action Research

Although not all textbooks agree, most generally divide the field of research into three categories: Basic (pure) research, applied research, and action research. **Basic research** is conducted mostly in labs, under tightly controlled conditions, and its main goal is to develop theories and generalities. This type of research is not aimed at solving immediate problems or at testing hypotheses. For example, scientists who worked in labs, using animals, such as mice and pigeons, developed the theory of behaviorism. These early behaviorists did not have an immediate application for their theory when it was first developed.

Applied research is aimed at testing theories and applying them to specific situations. Based on previously developed theories, hypotheses are then developed and tested in studies classified as applied research. For example, based on the theory of behaviorism, managers hypothesized that subordinates' behaviors will improve when incentives are provided. Next, studies were conducted where incentives, such as an increase in pay, were used as a reward for subordinates whose performance needed improvement. After the introduction of the increase in pay, the employees' behavior was monitored and assessed to determine the effectiveness of the intervention.

Action research, as the term implies, is undertaken to solve a problem by studying it, proposing solutions, implementing the solutions, and assessing the effectiveness of these solutions. The process of action research is cyclical; the researcher continues to identify a problem, propose a solution, implement the solution, and assess the outcomes.

Action research is the most common type of research used in business. Managers use research to generate information to help make decisions in all functional areas of business. People managing the accounting, finance, production, marketing, and all other tasks in business need information to make informed decisions. Action research is used to provide the information to help make those decisions. It is important to note that research supplements, but does not replace the experience and decision-making skills of managers in an organization.

Quantitative vs. Qualitative Research

Most research textbooks describe methods and approaches as either quantitative or qualitative. **Quantitative research** is defined as research that focuses on explaining cause-and-effect relationships, studies a small number of variables, and uses numerical data. Researchers conducting quantitative research usually maintain objectivity and detach themselves from the study environment. This research approach usually starts with a hypothesis and the study is designed to test this hypothesis. Quantitative researchers believe that findings can be generalized from one setting to other similar settings, and are looking for laws, patterns, and similarities. **Qualitative research** is defined in most textbooks as that which seeks to understand *why* things happen the way they do. Usually, in such research, the researcher focuses on one or a few cases, which are studied in depth. Qualitative research is more subjective in nature (e.g., interviews, observations, and focus groups). Qualitative research is context-based, recognizing the uniqueness of each individual and setting.

Quantitative research is *not* the same as experimental research, although a great deal of quantitative research is experimental. And, while it is true that qualitative research is descriptive, qualitative researchers also use numerical data, such as when they count events or perform certain data reduction analyses. These quantitative and qualitative paradigms are not a simple, clear way to classify research studies, because they are not two discrete sides of a coin. Rather, the paradigms are two end points on a continuum and studies can be located at different points along this continuum.

In the past, researchers identified themselves as either "qualitative researchers" or "quantitative researchers" and the two paradigms were seen as completely different from each other. Today, while recognizing the differences between the two paradigms, more and more researchers see the two as complementary and support using both in mixed-methods research studies.

In a typical quantitative study, data are collected to describe phenomena or to test hypotheses. Statistical techniques are then used to analyze the data. *This book, like most other business statistics textbooks, is geared toward the analysis of quantitative, numerical data.*

Experimental vs. Nonexperimental Research

The third way to classify research is to distinguish between experimental research and nonexperimental research.[1] In **experimental research**, researchers plan an intervention and study its effect on groups or individuals. The intervention is called the **independent variable** (or *treatment*), while the outcome measure is called the **dependent variable**. The dependent variable is used to assess the effectiveness of the intervention. For example, the independent variable may be a new employee training method, advertising campaign, or accounting procedure. Examples of dependent variables are employee performance, market share, level of consumer satisfaction, or employee motivation.

Nonexperimental research may be divided into two types: causal comparative and descriptive. **Causal comparative research** (also called **ex post facto**), like experimental research, is designed to study cause-and-effect relationships. However, unlike experimental research, in causal-comparative studies, the independent variable is not manipulated for two main reasons: Either it has occurred prior to the start of the study, or it is a variable that cannot be manipulated. **Descriptive research** is aimed at studying a phenomenon as it is occurring naturally, without any manipulation or intervention. This is used often in marketing surveys and many other types of business research. Researchers are attempting to describe and study phenomena, and are not investigating cause-and-effect relationships. Following is a discussion of experimental research, followed by a brief discussion of nonexperimental research.

---

1. **A HINT:** Nonexperimental research may also be called descriptive research.

Experimental Research

Most experimental studies are conducted to compare groups. In such studies, the experimental group members receive the treatment, e.g., new training method, while members in the control group either receive the traditional approach (i.e., old training method) or do not receive any treatment. An example might be a study conducted by a corporate trainer who wants to test the effectiveness of using an Internet website to enhance his training class. The trainer teaches the same class to two groups of trainees and is using an Internet website with one of the groups, but not with the other. The Internet website, established, moderated, and facilitated by the trainer, enables the trainer and the trainees to communicate with and amongst each other. The site contains materials prepared by the trainer, such as a course schedule or outline, handouts, assignments, review exercises, suggested reading, and other Internet sites to be used as resources. The Internet website may also be used by the trainees to discuss projects, pose questions, suggest activities, and more. The trainees in the other training class serve as the control group. They continue their class using the same techniques that were used in previous training sessions. At the end of the training sessions, the trainees from both groups take an examination created by the trainer. Test scores of the trainees in the experimental group who used the class Internet website are compared to the scores of the students in the control group to determine the effect of the Internet website on their performance in the training sessions.

In other cases, no treatment is applied to control-group members who are being compared to the experimental group. For example, when researchers want to study the effect of a new advertisement on buying behavior, they may recruit two groups of consumers; one group would watch the advertisement, while the other group of consumers would not see the new ad. Then, both groups of participants would be asked about their buying behavior relative to the product featured in the new advertisements to determine if the consumers who watched the new advertisements are more likely to buy the product than those who did not watch the new ads.

Other researchers, posing the same research question about the effect of new advertisements on consumers, may choose another experimental design and have these consumers serve as their own control. They may design a study where the behaviors of the same consumers would be studied twice: once before and once after they watch the new ads. Then, the researchers would note any change in behavior of the consumers in the group, all of whom were administered the treatment (i.e., watching the new ads).

As mentioned before, researchers conducting experimental research study the effect of the independent variable on the dependent variable. However, when researchers observe changes in the dependent variable, they have to confirm that these changes have occurred as a result of the independent variable and are not due to other variables, called extraneous variables. Extraneous variables are other plausible explanations that could have brought about the observed changes in the outcome variable. For example, suppose a company tries a new compensation method on half of its salespeople. If those to whom the new compensation is being provided suddenly sell more than those using the traditional method of compensating salespeople in the company, the researchers have to first confirm that the higher sales results are due to the new compensation method, rather than other extraneous variables. Those extraneous variables could be differing levels of skills and experience between the salespeople, different types of customers, or the economic conditions in the different territories in which the salespeople work. Prior to starting the study, the researchers have to review and control all possible extraneous variables that might affect the outcomes of the study. In our example, the researchers may want to ensure that both groups, experimental and control, are similar to each other before the new compensation method is implemented. The researchers have to document and verify that both groups have similar skills and experience levels and are selling similar products to similar customers. When the extraneous variables are controlled, it is assumed that the groups differ from each other on one variable only—the compensation method for the salespeople. If the experimental group of salespeople sells more, the researchers can conclude that the new compensation method is effective.

At times, extraneous variables develop during the study and are unforeseen. When researchers observe unexpected outcomes at the end of their study, they may want to probe and examine whether some unplanned extraneous variables are responsible for those outcomes. Often, when researchers fail to confirm the hypotheses stated at the beginning of their study, they examine the study to determine if any extraneous variables are responsible for these unexpected findings. When reporting their results, researchers are likely to include a discussion of possible extraneous variables in order to explain why their hypotheses were not confirmed.

A study is said to have a high **internal validity** when the researchers control the extraneous variables and the only obvious difference between the experimental and control groups is the intervention (i.e., the independent variable). It makes sense, then, that a well-designed experiment has to have high internal validity to be of value. When there are uncontrolled, extraneous variables, present competing explanations that can account for the observed changes in the dependent variable. One way to eliminate threats to internal validity and increase internal validity is to conduct studies in a lab-like setting under tight control of extraneous variables. Doing so, though, decreases the study's external validity. **External validity** refers to the extent to which the results of the study can be generalized and applied to other settings, populations, and groups. Clearly, if researchers want to contribute to their field (e.g., business) their studies should have high external validity. The problem is that when studies are tightly controlled in order to have *high* internal validity, they tend to have *low* external validity. Thus, researchers have to strike a balance between the two. First and foremost, every study should have internal validity. But, when researchers control the study's variables too much, the study deviates from real life, thus decreasing the likelihood that the results can be generalized and applied to other situations. Since experimental studies must have internal validity to be of value, a brief discussion of the major threat to internal validity is presented next.[2]

Threats to Internal Validity

1. *History* refers to events that happen while the study takes place that may affect the dependent variable. For example, suppose an economics professor wants to study the effectiveness of a new method for presenting material on fiscal policy (the independent variable) she is teaching to first year macroeconomics students. The dependent variable is the students' scores on a fiscal policy exam. The scores of students from the previous year are compared to those of this year's students who are exposed to the new method. However, during this year unemployment in the U.S. rises dramatically despite well publicized fiscal policy implemented by the Federal government. Assume, further, that this year's students score higher on the fiscal policy examination compared with last year's students. When the professor evaluates the effectiveness of the new method, she should consider the effect of history as a possible threat to the internal validity of her study. She should confirm that the higher scores on the fiscal policy examination are due to the planned intervention (i.e., the new method for presenting the material) and not to the unemployment and well-publicized, apparently inadequate fiscal policy that occurred while the study was going on.

2. *Maturation* refers to physical, intellectual, or mental changes experienced by participants while the study takes place. Maturation is a particular threat to internal validity in studies that last for a longer period of time (as opposed to short-duration studies), or in studies that involve younger people who experience rapid changes in their development within a short period of time.

3. *Testing* refers to the effect that a pretest has on the performance of people on the posttest. For example, in a study designed to test a new advertising campaign, subjects are asked how likely they are to buy the advertised product before and after they see the new advertising campaign. If they say they are more likely to buy on the posttest, it may be simply because they were asked about buying the product before, rather than due to the effectiveness of the new advertising.

2.   **A HINT:** For more information about threats to internal and external validity, you may want to review a book by Campbell and Stanley that is considered the most-cited source on the topic of experimental designs. The title of the book is: Campbell, D.T., & Stanley, J.C. (1971). *Experimental and quasi experimental designs for research.* Chicago: Rand McNally

4. *Instrumentation* refers to the level of reliability and validity of the instrument being used in the study. For example, in a research instrument designed to assess the effectiveness of a new financial model's ability to predict future profits, profit is used as the dependent variable. If the researcher finds that the profits of the companies studied are not what the researcher predicted, it may not be an indication that the model is ineffective. Rather, it may be that the research instrument used to evaluate the model does not measure the relationship between the independent and dependent variables. In other words, the test lacks in validity and does not measure the effectiveness of the new financial model.

5. *Statistical regression* refers to a phenomenon whereby people who obtain extreme scores on the pretest tend to score closer to the mean of their group upon subsequent testing, even when no intervention is involved. For example, suppose an IQ test is administered to a group of business students. A few weeks later, the same students are tested again, using the same test. If we examine the scores of those who scored at the extreme (either very high or very low) when the test was administered the first time, we would probably discover that many low-scoring students score higher the second time around, while many high-scoring students score lower. The statistical regression phenomenon may pose a threat to internal validity in certain studies where the following occur: Participants are selected for the study based on the fact that they have scored either very high or very low on the pretest and the participants' scores on the posttest serve as an indicator of the effectiveness of the intervention.

6. *Differential selection* may be a threat in studies where volunteers are used in the experimental groups or in studies where pre-existing groups are used as comparison groups (e.g., one serves as experimental group and one as control group). This phenomenon refers to instances where the groups being compared differ from each other on some important characteristics even before the study begins. For example, if the experimental group is comprised of volunteers, they may perform better because they are more interested in the study and are highly motivated, rather than as a result of the planned intervention.

Threats to External Validity

Threats to external validity may limit the extent to which the results of the study can be generalized and applied to populations that are not participating in a study. It is easy to see why people might behave differently when they are being studied and observed. For example, being pretested or simply tested several times during the study may affect people's performance and motivation. Another potential problem may arise in studies where both experimental and control groups are comprised of volunteers and, therefore, may not be representative of the general population. Other potential problems which may pose a threat to external validity include: (a) People may react to the personality or behavior of those who observe or interview them; (b) People may try harder and perform better when they are being observed or when they view the new intervention as positive even before the start of the study; and (c) Researchers may be inaccurate in their assessments when they have some prior knowledge of the study's participants before the start of the study.

Two well-known examples serve to illustrate some potential threats to external validity. One is called the **Hawthorne Effect**, named after a study conducted in a Western Electric Company plant in Hawthorne, Illinois, near Chicago. In this study, researchers wanted to assess the effect of light intensity on workers' productivity. When the researchers increased the light intensity, it resulted in an increase in productivity. To confirm that the change in light intensity was indeed the cause for the increased productivity, the researchers *decreased* the light intensity. They discovered that productivity still went up. This experiment led them to conclude that the reason productivity went up in first place was not the change in light intensity, but people's perception that they are being studied. Today, when designing experiments, researchers take into consideration the Hawthorne Effect, whereby the study's participants may behave in a certain way not necessarily because of the planned intervention but rather as a result of their knowledge that they are being observed and assessed.

Another threat is called the **John Henry Effect**. John Henry worked for a railroad company when the steam drill was introduced with the intention of replacing manual labor. John Henry, in his attempt to prove that men can do a better job than the steam drill, entered into a competition with the machine. He did win, but dropped dead at the finish line. Today, the John Henry Effect refers to conditions where control group members perceive themselves to be in competition with experimental group members and therefore perform above and beyond their usual level. In a study where the performance of control and experimental groups are compared, an accelerated level of performance of control group members may mask the true impact of the intervention.

Comparing Groups

In conducting experimental research, the effectiveness of the intervention (the independent variable) is assessed via the dependent variable (the measured variable). In all experimental studies, a posttest is used as a measure of the outcome, although not all studies include a pretest. Researchers try to compare groups that are as similar as possible prior to the start of the study, so that any differences observed on the posttest can be attributed to the intervention. One of the best ways to create two groups that are as similar as possible is by randomly assigning people to the groups. Groups that are formed by using random assignment are considered similar to each other, especially when the group size is not too small.[3] When the groups being compared are small, they are likely to differ from each other, even though they may have been created through random assignment. Also, keep in mind that in real life, researchers are rarely able to randomly assign people to groups and they often have to use existing intact groups in their research studies.

Another approach that is used by researchers to create two groups that are as similar as possible, is matching. In this approach, researchers first identify a variable which they believe may affect, or be related to, people's performance on the dependent variable (e.g., the posttest). Pairs of people with similar characteristics on that variable are randomly assigned to the groups being compared. For example, two people who have the same verbal score on the Graduate Management Admission Test (GMAT) may be assigned to the experimental or control group in a study that involves a task where verbal ability is important. The limitations of matching groups include the fact that the groups may still differ from each other because we cannot match them based on more than one or two variables. Also, there is a good possibility that we would end up with a smaller sample size because we would need to exclude from our study those people for whom we cannot find another person with the same matching score.

Experimental group designs can be divided into three categories: pre-experimental, quasi-experimental, and true experimental designs. The three categories differ from each other in their level of control and in the extent to which the extraneous variables pose a threat to the study's internal validity. In studies classified as pre-experimental and quasi-experimental, when groups are compared, researchers may still not be able to confirm that differences between the groups on the posttest are caused by the intervention. The reason is that these studies involve the comparison of intact groups, which may not be similar to each other at the beginning of the study. True experimental designs offer the best control of extraneous variables.

Studies classified as pre-experimental do not have a tight control of extraneous variables, thus their internal validity cannot be assured. Studies using pre-experimental designs either do not employ control groups or, when such groups are used, no pretest is administered. Thus, researchers cannot confirm that changes observed on the posttest are truly due to the intervention.

Studies using quasi-experimental designs have a better control than those in pre-experimental designs; however, there

---

3. **A HINT:** A rule of thumb by most researchers recommends a group size of at least 30 for studies where statistical tests are used to analyze numerical data.

are still threats to internal validity and potential extraneous variables are not well controlled in such studies. In quasi-experimental designs, the groups being compared are not assumed to be equivalent at the beginning of the study. Thus, any differences observed at the end of the study may not have been caused by the intervention, but are due to pre-existing differences. It is probably a good idea to acknowledge these possible pre-existing differences and to try to take them into consideration while designing and conducting the study, as well as when analyzing the data from the study. Using a pretest to document any possible pre-existing differences would help researchers assess gains and evaluate the differences between groups. Studies using quasi-experimental designs include time series and counterbalanced designs. In time-series designs, groups are tested repeatedly before and after the intervention. In counterbalanced designs, several interventions are tested simultaneously and the number of groups in the study equals the number of interventions. All the groups in the study receive all interventions, but in a different order.

In true experimental design, participants are randomly assigned to groups. Additionally, if at all possible, the study's participants are drawn at random from the larger population before being randomly assigned to their groups. Since the groups are considered similar to each other when the study begins, researchers can be fairly confident that any changes observed at the end of the study are due to the intervention.

Comparing Individuals

While most experimental studies involve groups, some studies in business focus on individuals. These studies' designs, where individuals are used as their own control, are called **single-case** (or **single-subject**) **designs**. In these studies, individuals' behavior or performance is assessed during two or more phases, alternating between phases with or without an intervention. The measure used in single-case studies (i.e., the dependent variable) is collected several times during each phase of the study, to ensure its stability and consistency. Since the measure is used to represent the target behavior, the number of times the target behavior is recorded in each phase may differ from one study to another. When diagramming these designs, the letter A is used to indicate the baseline phase where no intervention is applied, and the letter B is used to indicate the intervention phase.

One of the most common single-case designs involves three phases and is called the **A-B-A single-case design**. The study begins by collecting several measures of the target behavior to establish the baseline (phase A). Then an intervention is introduced, during which the same target behavior is again measured several times (phase B). Next, the intervention is withdrawn and the target behavior is assessed again (phase A). The target behavior is compared across all phases, *without* the intervention (phase A) and *with* the intervention (phase B), to determine if the intervention was effective. If the target behavior is improved during phase B, the researchers can speculate that this was caused by the intervention. However, to rule out any extraneous variables as the possible cause for the change in the target behavior, it is assessed again during the withdrawal phase (the third phase). The expectation is that since no intervention is used at this phase (the second phase A), the target behavior should return to its original level at the start of the study (phase A). However, if the target behavior improves even though the intervention has been withdrawn, we may speculate that the intervention has produced a long-term positive effect. Studies may include repeated cycles of the baseline (phase A), treatment (phase B), withdrawal of treatment (return to baseline A), and treatment (phase B). The basic single-case designs described above can be modified to include more than one individual and more than one intervention in the same study. Another modification to this design may be studies in which several individuals are studied simultaneously, and the length of time of the baseline and treatment phases (phases A and B) differs from one person to another.

There are several potential problems associated with single-case studies. Because only one or a few individuals are studied, the external validity of the study (the extent to which the results can be generalized to other populations and settings) may be limited. To overcome this problem, single-case studies should be replicated. Another problem involves the nature of the intervention that may be used in single-case studies. In certain cases, it is not possible for researchers to withdraw the intervention and return to the baseline phase. For example, in a study where the intervention includes the implementation of a new production technique, the manager may not be able to tell the subordinates during the withdrawal phase not to use the technique once they have mastered it in the intervention phase. In other cases, withdrawing the intervention may pose ethical dilemmas. If the results from phase B convince the researchers that the treatment is effective, they may be reluctant to withdraw it in order to return to the baseline phase.

Note that single-case designs that use quantitative data are different from case studies that are used extensively in qualitative research. In the latter type, one or several individuals or "cases" (such as an organization or department) are studied in depth, usually over an extended period of time. Researchers employing a qualitative case study approach typically use a number of data collection methods (such as interviews and observations) and collect data from multiple data sources. They study people in their natural environment and try not to interfere or alter the daily routine. Data collected from these nonexperimental studies are usually in a narrative form. In contrast, single-case studies, which use an experimental approach, collect mostly numerical data, and focus on the effect of a single independent variable (the intervention) on the dependent variable (the outcome measure).

Nonexperimental Research

As mentioned before, nonexperimental research may be divided into two categories: causal comparative (ex post facto) and descriptive. Causal comparative studies are designed to investigate cause-and-effect relationships, without manipulating the independent variable. Descriptive studies simply describe phenomena.

Causal-comparative (ex post facto) Research

In studies classified as causal comparative, researchers attempt to study cause-and-effect relationships. That is, they study the effect of the independent variable (the "cause") on the dependent variable (the "effect"). However, unlike in experimental research, the independent variable is not being manipulated because it has already occurred when the study is undertaken, or it cannot or should not be manipulated. The following examples are presented to illustrate these points.

Let's say a company wants to study the effect of education level on the performance of new management trainees. The independent variable (i.e., the trainee's education level before being hired by the company) occurred prior to the start of the study and therefore cannot be manipulated. Another example of causal-comparative research may be a study designed to assess the effect of consumers' gender on their attitudes toward a new shampoo product. Obviously, the independent variable (i.e., the consumers' gender) is predetermined and cannot be manipulated.

At times, the independent variable can be manipulated, but researchers would not do so for ethical reasons. For example, based on empirical data, researchers may speculate that teenagers who are exposed to beer advertising on television are more likely to drink beer before the age of 21. However, the relationship between children's exposure to beer advertising and their consumption of beer should not be studied using experimental design. For obvious reasons, researchers are not going to randomly select a group of teenagers and assign half of them to serve as the experimental group that is shown beer ads on television and then asked about their beer consumption.

Descriptive Research

Many business research studies are conducted to describe existing phenomena. Although researchers may construct new instruments and procedures to gather data, there is no planned intervention and no change in the routine of people or phenomena being studied. The researchers simply collect and interpret their study's data. Thus, it is easy to see why qualitative research is considered nonexperimental. Most researchers conducting nonexperimental research use qualitative approaches and collect narrative data, although nonexperimental researchers may use narrative or numerical data.

Quite often, researchers conducting descriptive research use questionnaires, surveys, and interviews. The census survey, for example, is a nonexperimental study. Other examples of findings from a nonexperimental study include information presented in a sales report—either that of an individual salesperson or that of an entire district or company. Information provided by governmental offices, such as the Consumer Price Index (CPI), is also based on nonexperimental research. Descriptive statistics may be used to analyze numerical data derived from nonexperimental studies. For example, an investment banking firm may compare the Return on Investment figures of the companies it represents.

Correlation is often used in descriptive research. In most studies using correlation, two or more variables are compared for similarities. (For a discussion of correlation, see Chapter 8.) For example, a marketing manager might want to correlate her advertising spending with her market share. If the correlation of the advertising expenditures and market share is high, the marketing manager may suggest conducting an experimental study whereby she increases spending more money on advertising to increase its market share.

When researchers want to study how individuals change and develop over time, they can conduct studies using *cross-sectional* or *longitudinal* designs. In **cross-sectional** studies, a subset of a population is studied at the same point in time. Each person in the different samples is studied one time only. The biggest advantage of this design is that it is saves time because data can be collected quickly. The biggest disadvantage is that we are studying different cohorts, rather than follow the same group of individuals.

Longitudinal studies are used to measure change over time by collecting data at two or more points for the same or similar groups of individuals. The greatest advantage of this design is that the same or similar individuals are being followed; however, a major disadvantage is that the study lasts a long time. There are three types of longitudinal studies: *Panel, cohort,* and *trend.*

In a **panel** study, the same people are studied at two or more points in time. For example, marketing research firms use panels to track changes in consumers' attitudes and behaviors. This information is used to change existing products and attempts to anticipate possible new products that might appeal to the changing needs of these consumers.

In a **cohort** study, similar people, selected from the same cohort, are studied at two or more points in time. For example, a university may survey its students to compare their attitudes towards the choice of classes offered to them. In the first year, a group of freshmen is selected and surveyed. In the second year, a group of sophomores is selected and surveyed. The following year, a group of juniors is selected and surveyed, followed by a group of seniors the next year.

In a **trend** study, the same research questions are posed at two or more points in time to similar individuals. For example, accountants may be asked about their opinion toward changes in the tax laws every year or every five years to allow researchers to record and note trends and changes over time.

Summary

1. **Research** is a systematic inquiry that includes data collection and analysis. The goal of research is to describe, explain, or predict present or future phenomena. These phenomena can occur in any function of business, including accounting, economics, finance, management, and marketing.

2. **Basic research**, whose main goal is to develop theories and generalizations, is conducted mostly in labs, under tightly controllesd conditions.

3. **Applied research** is aimed at testing theories and applying them to specific situations.

4. **Action research** is designed to solve a problem by proposing solutions to the problem, and assessing the effect of these solutions.

5. **Quantitative research** is conducted mostly to study cause-and-effect relationships and focuses on studying a small number of variables and collecting numerical data.

6. **Qualitative research** seeks to understand behavioral and social phenomena and focuses on one or a few cases that are studied in depth using multiple data sources.

7. Data sources used in qualitative research are subjective in nature (e.g., interviews and observations) and they yield mostly narrative data.

8. In many textbooks, quantitative research is equated with experimental designs and the use of numerical data, while qualitative research is equated with descriptive research and narrative data. A better way to describe quantitative and qualitative paradigms is to state that the paradigms are two end points on a continuum. Studies can be located at different points along this continuum.

9. In experimental research, researchers plan an **intervention** and study its effect on groups or individuals.

10. While most experimental studies use numerical data and most descriptive studies use narrative data, both numerical and narrative data can be used in experimental or descriptive studies.

11. In experimental research, researchers plan an **intervention** and study its effect on groups or individuals. The intervention is also called **the independent variable** or **treatment**.

12. Experimental research is designed to test the effect of the independent variable on the outcome measure, called the **dependent variable**.

13. **Nonexperimental research** may be divided into *causal comparative* (also called *ex post facto*) research and *descriptive* research.

14. Qualitative research is considered nonexperimental, and many researchers conducting nonexperimental research use qualitative approaches and collect narrative data.

15. Researchers conducting descriptive research may collect narrative or numerical data. The census is an example of a descriptive study where a large amount of numerical data is collected.

16. **Descriptive research** is aimed at studying phenomena as they occur naturally without any intervention or manipulation of variables.

17. In many experimental studies, the experimental group that receives the treatment (the intervention) is compared to the control group that receives no treatment or is using the existing method. In other experimental studies, the performance of the same group is compared before the intervention (the pretest) and after the intervention (the posttest).

18. **Extraneous variables** are variables—other than the planned intervention—which could have brought about changes that are measured by the dependent variable. Extraneous variables may be unforeseen and develop during the study, especially when the study lasts for a long period of time.

19. A study is said to have high **internal validity** when the extraneous variables are controlled by the researchers and the only obvious difference between the experimental and control groups is the planned intervention (i.e., the independent variable).

20. **External validity** refers to the extent to which the results of the study can be generalized and applied to other settings, populations, or groups.

21. Threats to internal validity include: **History, Maturation, Testing, Instrumentation, Statistical Regression,** and **Differential Selection.**

22. Threats to external validity include the **Hawthorne Effect** and the **John Henry Effect**.

23. Experimental group designs can be described as *pre-experimental*, *quasi-experimental*, and *true experimental*.

24. **Pre-experimental designs** do not have a tight control of extraneous variables and their internal validity is low.

25. In **quasi-experimental designs**, the groups being compared are not assumed to be equivalent prior to the start of the study. Studies in this category have a better control of extraneous variables compared with pre-experimental designs.

26. In **time-series designs**, which are considered quasi-experimental, groups are tested repeatedly before and after the intervention.

27. In **counterbalanced designs**, which are also considered quasi-experimental, several interventions are tested simultaneously and the number of groups in the study equals the number of interventions. All the groups in the study receive all interventions, but in a different order.

28. The most important aspect of **true experimental designs** is that participants are assigned at random to groups. Therefore, the groups are considered similar to each other at the start of the study.

29. In **single-case** (or **single-subject**) **designs**, the behavior or performance of people is assessed during two or more phases, alternating between phases with and without intervention. The measure (i.e., the dependent variable) used in a single-case study is administered several times during each phase of the study to ensure the stability and consistency of the data.

30. One of the most common single-case designs is the **A-B-A design**. In studies using this design, the target behavior (i.e., the dependent variable) is measured before the intervention (*phase A*), during the intervention (*phase B*), and during the second *phase A* when the intervention is withdrawn.

31. To increase the inherently low external validity of single-case studies where only one or a few individuals are studied, it is recommended that these studies be replicated.

32. **Nonexperimental research** is divided into two major categories: *causal comparative* (also called *ex post facto*) and *descriptive*.

33. When researchers want to study how individuals change and develop over time, they may use *cross-sectional* or *longitudinal* designs.

34. In **cross-sectional** studies, a subset of a population is studied at the same point in time.

35. **Longitudinal** studies are used to measure change over time by collecting data at two or more points for the same or similar groups of individuals over a period of time. There are three types of longitudinal studies: *Panel, cohort,* and *trend*.

36. In a **panel** study, the same people are studied at two or more points in time. In a **cohort** study, similar people, selected from the same cohort, are studied at two or more points in time. In a **trend** study, the same research questions are posed to similar individuals at two or more points in time.

Chapter 2

Basic Concepts in Statistics

The term **statistics** refers to methods and techniques used for describing, organizing, analyzing, and interpreting numerical data. Statistics are used by researchers and practitioners who conduct research in order to describe phenomena, find solutions to problems, and answer research questions.

Variables and Measurement Scales

A **variable** is a measured characteristic that can assume different values or levels. Some examples of variables are age, income level, gender, sales, and price of a product. By contrast, a measure that has only one value is called a **constant**. For example, a dollar is always equal to 100 cents and the number of hours in a day—24, is also a constant.

The decision whether a certain measure is a constant or a variable may depend on the purpose and design of the study. For example, the Gross Domestic Product (GDP) of different cities may be a variable in a study where several cities are included in an attempt to measure the economic growth of these cities. On the other hand, in a study of the growth of businesses within the same city, the GDP of the city is a constant.

A variable may be continuous or discrete. **Continuous variables** can take on a wide range of values and contain an infinite number of small increments. Time, for example, is a continuous variable. Although we may use increments of 1 minute or 1 second, it could take a fraction of a second for a computer to make a calculation. **Discrete variables**, on the other hand, contain a finite number of distinct values between any two given points. Some test scores are a discrete variable. For example, on an accounting test, a student may get a score of 20 or 21, but not a score of 20.5.

It is important to remember that in the case of intelligence tests, for instance, while the test can only record specific scores (discrete variable), intelligence itself is a continuous variable. On the other hand, research reports may describe discrete variables in the manner usually prescribed for continuous variables. For example, in reporting the number of employees working at each location of a retail store, a study may indicate that there are, on average, 26.4 employees per location. In actuality, the number of employees in any given location would, of necessity, be indicated by a whole number (e.g., 26 or 27). Reporting the discrete variable in this manner lets the researcher make finer distinctions, allowing for more sensitivity to the data than would be possible if adhering to the format generally used for the reporting of discrete variables.

Measurement is defined as assigning numbers to observations according to certain rules. Measurement may refer, for example, to counting the number of times a certain phenomenon occurs or the number of people who responded "yes" to a question on a survey. Other examples include using tests to assess personality or measure employee or consumer attitudes. Each system of measuring uses its own units to quantify what is being measured (e.g., dollars, percentiles, frequencies, and attitudes).

There are four commonly used types of measurement scales: *nominal, ordinal, interval*, and *ratio*. For all four scales we use numbers, but the numbers in each scale have different properties and should be manipulated differently. It is the duty of the researcher to ascertain the scale of the numbers used to quantify the observations in order to determine the appropriate statistical test that should be applied to analyze the data.

Nominal Scale

In **nominal scales**, numbers are used to label, classify, or categorize data. For example, the numbers assigned to the members of a football team comprise a nominal scale, where each number represents a player. Numbers may also be used to describe a group in which all members have some characteristic in common. For example, in coding data from a survey to facilitate computer analysis, males may be coded as "1" and females as "2." In this instance, it clearly does not make sense to add or divide the numbers. We cannot say that two males, each coded as "1," equal one female, coded as "2," although in other contexts, 1+1=2. Similarly, it will not make sense to report that the average gender value is, for example, 1.5! For nominal scales, the numbers are assigned arbitrarily, and are interchangeable. Consequently, instead of assigning 1 to males and 2 to females, we can just as easily reverse this assignment, and code males as 2 and females as 1.

Ordinal Scale

In **ordinal scales**, the observations are ordered based on their magnitude or size. This scale has the concept of *less than* or *more than*. For example, responses on a survey administered to college students may be used to rank-order their favorite fast-food establishments. Based on results of the survey, we may know that a particular fast food restaurant that is ranked as the 3rd most popular by students in a survey is preferred by these students over another fast food restaurant is that is ranked 5th by the same students. However, we do not know how many points separate the two restaurants. The same can be said about three medal winners in a long jump at the Olympic Games. It is clear that the gold medalist performed better than the silver medalist, who, in turn, did better than the bronze medalist. But we should not assume that the same number of inches separate the gold medalist from the silver medalist, as those inches separating the silver medalist from the bronze medalist. Thus, in an ordinal scale, observations can be rank-ordered based on some criterion, but the intervals between the various observations are not assumed to be equal.

Interval Scale

Interval scales have the same properties as ordinal scales, but they also have equal intervals between the points of the scale. Most of the numerical examples used in this book are measured using an interval scale. Temperatures, calendar years, IQ, and attitude survey scores, all are considered interval scales. The difference between a temperature of $20°F$ and $25°F$ is $5°F$, and is the same as, let's say, the difference between $65°F$ and $70°F$. However, we cannot say that a temperature of $90°F$ is three times as hot as a temperature of $30°F$; or that an individual with an IQ of 150 is twice as smart as one with an IQ of 75, because an interval scale does not have an absolute, or true, zero. An absolute zero is a point lacking completely the characteristic being measured. In Fahrenheit temperature, the temperature of $0°$ does not imply lack of heat. (The absolute zero is $-273°F$, where the molecules do not move at all.) Similarly, the zero point in an IQ scale is not a true zero, because we cannot say that a person who received a score of zero on our IQ test has no intelligence at all. We probably can find other questions that this person can answer, but these questions were not asked on the IQ test.

Ratio Scale

Ratio scales have the same characteristics as interval scales, but in addition they have an absolute zero. Thus, we can compare two points on the scale and make statements such as: this point is twice as high as that point, or this person is working half time (as opposed to full time). Money, for example, is measured on a ratio scale. We can say that a company that has earnings of $100,000 is half as profitable as a company that earns a profit of $200,000. Money has a true zero point that can be seen in a checkbook, sometimes right before payday. Or, for example, in a race, the absolute zero point is when the gun sounds and the stopwatch is pressed to start the counting. Ratio scales exist most often in the physical sciences but less frequently in business and other social sciences.

Populations and Samples

An entire group of persons or elements that have at least one characteristic in common is called a **population**. Examples would include all the residents of a particular suburb, all the economists that work for the Federal Reserve, or all the states in the U.S. A population may have more than one characteristic, or trait, in common. For example, we may talk about a population of female students at the local state university who are majoring in computer sciences.

In real life, rarely do we study and measure entire populations. The most notable example of a study of the entire population is that of the census, which is conducted once every 10 years. Clearly, including *all* members of a population in a study is expensive, time consuming, and simply impractical. Yet, most research studies are concerned with generalization and obtaining rules and findings that describe large groups. Thus, quite often, the researchers draw a sample and use it to gain information about the population. A **sample**, then, is a small group of observations selected from the total population. A sample should be *representative* of the population, because information gained from the sample is used to estimate and predict the characteristics of the population that are of interest.

As an example, suppose we want to know what Certified Public Accountants (CPAs) in six Midwestern states think about the Sarbanes-Oxley Act. There are thousands of CPAs in these states and it is too expensive to survey every one of them. Instead, a sample of 500 CPAs may be selected and surveyed. The results of this survey can be said to represent all the CPAs in the six states. Of course, as with every survey, the response rate has to be adequate to assure that the results truly reflect the total population.

A certain level of error is expected when we use samples to estimate populations. Some chance variation in the sample numerical values (e.g., mean) occur when we repeatedly select same-size samples from the same population and compare their numerical values. This error, called a **sampling error**, is beyond the control of the researcher.

Parameters and Statistics

A measure that describes a characteristic of an entire population is called a **parameter**. The total number of workers with school-age children who work for a certain automobile manufacturing company is a parameter, because it describes a certain characteristic of the entire workforce in that company. A **statistic** is a measure that describes a characteristic of a sample.[1] For example, if we were to select at random a group of 500 workers from the same

---

1.   **A HINT:** Note the difference between two similar terms: **statistics** and a **statistic**. *Statistics* refers to a group of methods and techniques, whereas a *statistic* is a numerical index or value of a sample.

company and determine the number of workers who have school-age children, that number would be a statistic.

In most research studies, we are interested in obtaining information about a population parameter, but instead of obtaining the parameter directly by measuring every member of the population, we draw a sample, measure the sample to obtain the statistic, and then use that statistic to estimate the corresponding population parameter.
For example, an electronics company that develops computer action games may want to pilot test a new game designed to be marketed to teenage children. The company may select a sample of teenage children, let them play with the new computer game, observe their reactions to the game, and ask them for their opinions. The company will then generalize the findings from the sample to the total population of teenagers who are the potential users of this new computer game.

The majority of the research studies in business are designed to study populations by using samples that are representative of these populations. In many physical sciences studies it is quite simple to obtain a representative sample. For example, assume that a scientist wants to study the quality of the water of a pool. All the scientist needs to do is to scoop out a small jar with the pool water and analyze this sample. We would all agree that the sample of water in the jar is representative of the pool water. In business, as well as in other behavioral sciences (e.g., psychology and sociology), the task of obtaining a representative sample of employees, consumers, products, or financial information, can be more complicated.

There are several sampling methods that may be used in research. In choosing the sampling method, the researcher has to decide which one is appropriate and feasible in a given situation. Most sampling methods share the same steps, or sequence: First, the population is identified, then the sample size required is determined, and lastly, the sample is selected. The next section describes several of the most common sampling techniques that are used by researchers.

Methods of Sampling

Simple Random Sample

In selecting a **simple random sample**, every member of the population has an equal and independent chance of being selected for inclusion in the sample. That is the selection of one member in no way increases or decreases the chances of another member to also be selected. Sampling procedures whereby the first 100 people who stand in line are chosen, or every other person from a list is selected, do not fit the definition of a random sample. When the first 100 people are selected, those who stand behind them do not have a chance of being included in the sample. Likewise, choosing every other person means that persons next to those being selected do not have a chance of being included.

In theory, if the random sample is large enough, it will truly represent the population in every respect, and be an accurate reflection of the population. On the other hand, selecting 10 people from a population of 1000, even if done by using a random sampling procedure, may result in a sample that is not truly representative of the population.

The typical idea that comes to mind when we think of a random sample is drawing names out of a hat. While a random sample can be drawn this way, such a process is not efficient. There are more practical means of achieving the same results, using, what we call a table of random numbers which can be found in many statistics textbooks. However, most researchers are likely to use a computer program to select random samples for their studies.

Systematic Sample

In a **systematic sample**, every *Kth* member (e.g., every 5th or 10th person) is selected from a list of all population members. The procedure starts by ascertaining the size of the population and the desired sample size. The population size is then divided by the sample size to obtain the value of *K*. For example, if we have a population of 500 and need a sample of 25, we divide 500 by 25, to obtain a *K* of 20. In other words, we select every 20th member of the population to achieve the desired sample size of 25.

A systematic sample can be a good representation of the population when the names on the list from which the sample members are selected are listed randomly. Since this is rarely the case, the sample may be biased. For example, certain nationalities tend to have many last names that start with the same letter. Thus, a whole group of people of the same nationality may be excluded from the sample if the names of those selected are listed just before and just after that group of names.

Stratified Sample

To obtain a **stratified sample**, the population is first divided into subgroups (i.e., *strata*), and then a random sample is taken from each subgroup. Stratified sampling is used extensively in market research, in political polls, and in norming standardized tests. The final sample represents, proportionately, the various subgroups in the population. A stratified sample may be used when there is a reason to believe that various subgroups in the population may have different opinions or behave differently because of some characteristics that the group members have in common. An example may help to illustrate this sampling procedure.

Suppose a computing industry group with 5,000 members wants to survey its members about their attitudes toward merit pay. Instead of surveying all 5,000 members, a stratified sample of 250 may be selected. The members may first be divided into strata based on variables such as the type of organization for which work (business, government, non-profit), job title, and annual salary (in increments of $5,000). For example, we may have a stratum of members who work in government agencies as software programmers, whose annual salary is $60,000-$65,000. From each subgroup, a random sample may be drawn. The resulting sample of 250 members will include, proportionally, all subgroups from the computing industry population of 5,000 members. Thus, the sample of members that will participate in the survey will be a miniature of the population where each stratum is represented in proportion to its percent in the population.

Convenience Sample

Occasionally, researchers conduct studies using an accessible sample, such as the researchers' own business contacts, friends, or even family members. A **convenience** (or **incidental**) **sample** is a group that is chosen by the researcher to participate in the study because of its convenience. For example, college statistics professors may choose, for convenience and cost-cutting purposes, to use their own students to conduct an experimental study. Or graduate students working on their theses or dissertations may use their schools workplace to collect data. In fact, a fair number of research studies in business and other social sciences are done using an available, convenience sample.

The main problem in using an incidental sampling is that it is not always clear which population the sample represents, since the study did not start by choosing a sample from a defined population. Great care should be exercised in generalizing the results of the study to a larger population.

Sample Bias

Sample bias refers to *systematic* (as opposed to *random*) differences between the sample and the population from which it was selected. A biased sample contains a certain systematic error. If, for example, a political poll is conducted by randomly selecting respondents from the telephone book, the resulting sample is likely to be biased because it excludes voters with unlisted telephone numbers, or voters who do not have telephones.

A well-publicized example of such sample bias occurred during the 1936 presidential elections when Republican Alf Landon ran against Democrat Franklin D. Roosevelt. The *Literary Digest* predicted a victory by Landon after receiving a 25% response rate from 10 million written ballots, which were mailed out. The mailing list for the ballots was based on telephone books and state registries of motor vehicles. Of course, such a list left out a big segment of the population that voted in the presidential election, but was not included in the survey because they did not own a telephone or a car. As we all know, contrary to the prediction in the *Literary Digest*, Roosevelt defeated Landon in that race.

Another possible sample bias stems from using volunteers in a study. Even though the volunteers may come from a clearly defined population, they may not be "typical" of the other members of that population. Conducting a study with volunteers and then generalizing the results to the population at large can lead to incorrect conclusions.

A sample may also be biased when it is based solely on the responses of people who had mailed back their completed surveys. Those responding are often similar to people who volunteer to participate in a study. Therefore, their responses may, or may not, represent the rest of the population. In many cases, those who respond to surveys feel strongly one way or another about the topic of the survey, whereas the majority of people do not bother to respond. Yet, quite often, the responses of those who returned their surveys are generalized to the total population.

Size of Sample

As sample size increases, it is more likely to be representative of the population, especially when the sample is randomly selected. In well-designed experimental studies, as the population gets larger, a sample that is a smaller portion of the population may be sufficient to provide an accurate representation. When the population is greater than 10,000, a sample of 1,000 may give adequate precision.

In determining whether a sample truly represents the population, it is important to consider how the sample was selected, as well as the size of the sample used. For example, a sample that is drawn by using a simple random sampling approach is highly regarded. However, if that sample consists of five subjects only, it is probably not an adequate representation of the population from which it was selected. By the same token, size alone does not guarantee an accurate sample, and a large sample may also be biased. In general, it is recommended that researchers should try to obtain as large a sample as is feasible. A sample size of at least 30 cases or subjects is recommended in most studies in business.

Parametric and Nonparametric Statistics

There are different research situations that call for the use of two types of statistics: parametric and nonparametric.[2]

---

2.   **A HINT**: Nonparametric statistics are also called **assumption-free** or **distribution-free** statistics.

Parametric statistics are applied to data from populations that meet the following assumptions: the variables being studied are measured on an interval or a ratio scale; subjects are randomly assigned to groups; the scores are normally distributed; and the variances of the groups being compared are similar. When these assumptions are being met, researchers are likely to use parametric tests that are more efficient and powerful than their nonparametric counterparts. However, in many research situations in behavioral science and business, it is hard to meet all the required assumptions. As a result, findings should be interpreted cautiously. It is probably safe to say that researchers always use interval or ratio scales when applying parametric tests, while it is more common for researchers to violate the other assumptions.

Nonparametric statistics are used with ordinal and nominal data, or with interval and ratio scale data that fail to meet the assumptions needed for parametric statistics. Nonparametric statistics are easier to compute and understand, compared with parametric statistics. The chi square test, for example (see Chapter 12), is a nonparametric statistic, whereas the *t* test (see Chapter 10) and analysis of variance (see Chapter 11) are examples of parametric statistics. The majority of the statistical tests you are likely to read about in the literature are classified as parametric.

Descriptive and Inferential Statistics

The field of statistics is often divided into two broad categories: descriptive statistics and inferential statistics. **Descriptive statistics** classify, organize, and summarize numerical data about a particular group of observations. There is no attempt to generalize these statistics, which describe only one group, to other samples or populations. Some examples of descriptive statistics are the mean number of units produced in a factory each day, the number of consumers who buy a product, and the ethnic make-up of students at a given university.

Inferential statistics (which may also be called **sampling statistics**), involve selecting a sample from a defined population and studying that sample in order to draw conclusions and make inferences about the population. The sample that is selected is used to obtain sample *statistics*, which are used to estimate the population *parameters*. The rationale behind inferential statistics is that since the sample represents the population, what holds true for the sample probably also holds true for the population from which the sample was drawn.

In political polls, for example, a pollster may survey 1,500 voters and use their responses to predict the national election results the next day. Another example may be of a product manager who is conducting a test market study to explore the efficacy of different package designs on consumers' attitudes toward the product and likelihood of buying the product. Four different package designs were used at different times in the same supermarket during the course of one month. At the end of the month, the product manager compares survey results and the sales outcomes for the different package designs. These results are then used to decide which package design is most likely to maximize the sales of the product.

Descriptive and inferential statistics are not mutually exclusive. In a sense, inferential statistics include descriptive statistics. When a sample is observed and measured, we obtain descriptive statistics for that sample. However, inferential statistics can take the process one step further and use the information obtained from the sample to estimate and describe the population to which the sample belongs. Whether or not a given statistic is descriptive or inferential does not necessarily depend on the type of statistic, but rather on its purpose. For example, a one-day mean sales figure for a restaurant is a descriptive statistic if the restaurant manager wants to compare the sales from location of the restaurant to the sales results from another location. However, the manager may want to test new menu items at one of the restaurant's locations. In that study, sales from different locations of the restaurant chain are compared: the first location has changed its menu to include more Italian food and the other locations have not changed their menus. The results from the restaurant that changed its menu can be compared to 10 randomly selected locations that have not added Italian food to their menus to determine if, indeed, the menu change is attracting more business. If this were the case, the manager might consider adding Italian food to the other restaurant's menu.

Using Hypotheses in Research

A research study often begins with a **hypothesis** (an "educated guess") that is a prediction about the outcome of the study. After the hypothesis is proposed, a study is designed to test that hypothesis. The data collected in the study enable the researchers to decide whether the hypothesis is supported. Hypotheses should be clearly and concisely stated and be testable.

A study may have more than one hypothesis. For example, a study may be conducted to investigate gender differences in attitudes toward football, as well as the relationship between attitudes toward football and the number of hours spent watching football on television. Therefore, one hypothesis in this study may predict that the men's mean score on a questionnaire measuring attitudes toward football would be significantly higher than the women's mean score on that questionnaire. Another hypothesis may predict a positive correlation between scores on the questionnaire measuring attitudes toward football and the number of hours spent watching football on television by both men and women.

Alternative and Null Hypotheses

Two types of hypotheses are used in research to explain phenomena and to make predictions about relationships between variables in a study. These two hypotheses are the alternative hypothesis and the null hypothesis. The **alternative hypothesis** (represented by H_A or H_1) guides the investigation and gives direction to the design of the study. Often, the alternative hypothesis is simply referred to as *the hypothesis*.[3] It predicts that there would be some relationship between variables or difference between means or groups. For example, the alternative hypothesis may state that there would be a positive correlation between outdoor temperature and the sale of ice cream at an ice cream store. Or, an alternative hypothesis may predict that real estate agents using a new innovative real estate listing software program would list significantly more homes, compared with real estate agents not using the new software.

The **null hypothesis** (represented by H_0) predicts that there would be no relationship between variables or no difference between groups or means beyond that which may be attributed to chance alone. In most cases, the null hypothesis (which may also be called the **statistical hypothesis**) is not formally stated, but it is always implied. The following two examples may illustrate how the null hypothesis is used in business research. In the first example, we would conduct an experimental study to test the effect of payment method on sales. One retail location would allow customers to pay for purchases using any payment method they choose, including credit cards, while at another location of the same retailer, customers would not be allowed to pay for purchases with credit cards. The null hypothesis in this study states that there would be no difference in the sales results between the store accepting credit cards and the store not accepting credit cards. In our second example, the null hypothesis states that there would be no significant correlation between IQ and depression scores in college students. This hypothesis would be tested using two random samples of 200 students from two universities. IQ and depression scores of those students would be obtained and correlated to test the null hypothesis.

---

3.   **A HINT**: In this book, whenever we use the word *hypothesis*, we are referring to the *alternative hypothesis*, whereas the *null* hypothesis is always called *null hypothesis*.

Directional and Nondirectional Hypotheses

Hypotheses may be stated as *directional* or *nondirectional*. A **directional hypothesis** predicts the direction of the outcomes of the study. In studies where group differences are investigated, a directional hypothesis may predict which group's mean would be higher. In most experimental studies, when the hypothesis predicts differences in behavior or performance between the experimental and control group on the dependent variable, researchers are likely to use a directional hypothesis. In other words, they are quite certain that there would be a difference between the groups as a result of the intervention. In studies that investigate relationships between variables, directional hypotheses may predict whether the correlation will be positive or negative.

A **nondirectional hypothesis** predicts that there would be a difference or relationship but the direction of the difference or association is not specified. For example, the researcher predicts that one group's mean would be higher, but does not specify which of the two means would be higher. Similarly, when the researcher predicts a statistically significant relationship, but cannot predict whether the relationship would be positive or negative, the hypothesis is nondirectional.

Probability and Level of Significance

Statistical results from research studies may be used to decide whether to retain (i.e., accept) or reject the null hypothesis. Once this first decision is made, the researcher can then determine whether the alternative hypothesis has been confirmed. It should be mentioned, however, that this statistical decision is made in terms of *probability*, not *certainty*. We cannot *prove* anything; only describe the probability of obtaining these results due to sampling error or chance. For example, we may want to compare the means from experimental and control groups, using a statistical procedure called the *t* test for independent sample (see Chapter 10). The null hypothesis states that the difference between the two means is zero. The statistical results may lead us to two possible conclusions:

1. It is unlikely that the two means came from the same population, and the difference between them is too great to have happened by chance alone. The experimental treatment had a measurable effect, beyond chance, on the dependent variable. The null hypothesis is *rejected*.
2. The difference between the two means is not really greater than zero, and the two means probably did come from the same population. In such cases, even if we observe some differences between the two means, we attribute them to sampling error, and not to some systematic differences resulting from the experimental treatment. The null hypothesis is *retained*.

Regardless of the research hypothesis presented at the onset of the study, the statistical testing and the evaluation of the findings start with a decision regarding the *null* hypothesis. To make a decision about the null hypothesis, we first calculate the sample statistic to get the *obtained* value. We then compare the obtained value to the appropriate *critical* value, which is often determined from statistical tables. If the obtained value *exceeds* the critical value, the null hypothesis is *rejected*. *Rejecting* the null hypothesis means that the probability of obtaining these results by chance alone is very small (e.g., 5% or 1%). We conclude that the relationship or difference, as predicted by the alternative hypothesis (H_A), is probably true. *Retaining* the null hypothesis means that these results (e.g., difference between two means) may be due to sampling error and could have happened by chance alone more than 5% of the time.

In most statistical tests, the probability level of 5% (*p* value of .05) serves as the cut-off point between results considered **statistically significant** and those considered **not statistically significant**.[4] The *p* level (i.e., **level of**

---

4. **A HINT**: The term *significant* does not necessarily mean the same as "useful in practice" or "important."

significance) indicates the *probability* that we are rejecting a true null hypothesis. Findings are usually reported as *statistically significant* if the probability level is 5% or less ($p \leq .05$).[5] If the probability level is higher than 5% ($p > .05$), many researchers are likely to report the findings as *not statistically significant*, rather than report the *actual p* level. However, as you read published research reports, you may find that researchers often list the exact probability level (*p* value) instead of using the 5% cut-off point.

There is a clear relationship between the sample size and the confidence level in rejecting the study's null hypothesis. As the sample size *increases*, a *lower* computed test statistic value is needed in order to reject the null hypothesis at the $p=.05$ level. To illustrate this point, let's look at studies that use the Pearson correlation. The null hypothesis in such studies is that the correlation coefficient *r* is equal to zero ($r=0$). (See Chapter 8 for a discussion of correlation.) For example, with a sample size of 30 (n=30), the correlation coefficient has to be at least .349 ($r=.349$) to be considered statistically significant at $p=.05$. As the sample size increases to 50 (n=50), a correlation coefficient of $r=.273$ would be considered statistically significant at $p=.05$. And, and when the sample size is 100 (n=100), a correlation coefficient as low as $r=.195$ is statistically significant at $p=.05$.

Note the inverse relationship between the sizes of the samples and the magnitude of the correlation coefficients. Thus, with a very large sample size, even very low correlation coefficients are going to be defined as statistically significant. Conversely, with a very small sample size, even very high correlation coefficients are not going to be considered statistically significant.

This book provides step-by-step explanations of the processes for determining the *p* values in each of the statistical tests examples in the book (see Chapters 8-12). However, in real life you will probably use a computer program to compute the appropriate *p* values. There are several powerful computer software programs (such as SPSS) readily available to novices, as well as experienced researchers. These programs can analyze your statistical data and will, in most cases, provide the exact *p* values.

Errors in Decision Making

When the probability level is set at the beginning of the study, *before* collecting and analyzing the data, it is represented by the Greek letter **alpha (α)**. The convention is to use an alpha of .05. However, in some exploratory studies, researchers may set alpha at .10. In other studies, the researchers may want to set the alpha level at .01, so as to have a higher level of confidence in their decision to reject the null hypothesis.

When researchers decide to *reject* the null hypothesis (H_0) when in fact it is true and should not be rejected, they are making a **Type I error**. And, when they decide to *retain* the null hypothesis when in fact it should be rejected, they are making a **Type II error**. The proper decision is made when researchers reject a false null hypothesis, or when they retain a true null hypothesis.

If we decide to set alpha at $p=.01$ (instead of .05), we *decrease* the chance of making a Type I error because we are less likely to reject the null hypothesis. However, in setting alpha at .01, we *increase* the chance of making a Type II error and are more likely to retain the null hypothesis when in fact we should have rejected it.

---

5.   **A HINT**: When the results are statistically significant, report the highest (the best) level of significance. For example, if results are significant at the $p < .01$ level, report that level, rather than $p < .05$. Of course, you can always report the exact p value (e.g., $p = .03$).

Degrees of Freedom

In order to consult tables of critical values (found usually in the appendix of statistical textbooks), the researcher needs to know the **degrees of freedom** (*df*). Essentially, *df* is **n-1** (the number of cases or subjects in the study, minus 1), although there are some modifications to this rule in some statistical tests (e.g., the chi square test; see Chapter 12). The exact way to calculate the *df* will be explained in the discussion of each of the statistical tests that are included in this book.

Effect Size

The decision whether to retain or reject the null hypothesis is affected greatly by the study's sample size. A large sample size may lead researchers to reject the null hypothesis even when there are very small differences between the variables or when the correlation between the variables is very low. Conversely, in studies where a small sample size is used, researchers may decide to retain the null hypothesis even when there are large differences or high correlation between variables. In the last decade, the concept of *effect size* has gained much popularity as another way to evaluate the statistical data gathered in research studies. The American Psychological Association (2001) recommended the inclusion of effect size in the *Results* sections of research reports where the numerical results of studies are presented.[6]

Effect size (abbreviated as **ES**) is an index that is used to express the strength or magnitude of a difference between two means. It can also be used to indicate the strength of an association between two variables using correlation coefficients. Effect size is scale-free and can be used to compare outcomes from different studies where different measures are used. Effect size is not sensitive to sample size and can be computed regardless of whether the results are statistically significant. Using effect size in addition to tests of statistical significance allows researchers to evaluate the *practical* significance of the study's finding. There are several ways to calculate effect sizes, but one of the most commonly used approaches is the index called *d*, which was developed by Cohen (1988).[7]

Effect size can be used to compare the performance of two groups, such as experimental and control groups, or women and men. It can also be used in experimental studies where pretest and posttest mean scores are being compared. The comparison of the means is done by converting the difference between the means of the groups into standard deviation units.

When interpreting statistical results, researchers should look at the direction of the outcome (e.g., which mean is higher or whether the correlation is positive or negative) and whether the test statistics they compute are statistically significant. When appropriate, the effect size should also be computed to help researchers evaluate the practical importance of their data. (See Chapter 10 which has two examples where the effect size is used in interpreting statistical data obtained using the *t* test.)

Once the effect size is calculated, it can then be evaluated and interpreted. While no clear-cut guidelines are available to interpret the magnitude of the effect size, many researchers in behavioral sciences and education follow guidelines suggested by Cohen (1988). According to Cohen, an effect size (i.e., *d*) of .2 is considered small; an effect size of .5 is considered medium; and an effect size of .8 is considered large. Guidelines developed by other researchers define effect size of .2 as small; effect size of .6 is moderate; effect size of 1.2 is large; and effect size of 2.0 is very large. Effect sizes of 1.00 or higher are rare in business and other types of social science research.

---

6.   **A HINT:** See American Psychological Association (2001). *Publication manual of the American Psychological Association* (5th ed.). Washington, DC: Author.

7.   **A HINT:** See Cohen, J. (1988) *Statistical power analysis for behavioral sciences* (2nd ed.). Hillsdale, NJ: Lawrence Erlbaum.

Whether an effect size is considered practically significant may depend not only on its magnitude but also on the purpose and expectation of the researcher and the type of data being collected. For example, to change business practices that cost a great deal of money, time, and resources, an effect size of .7 may not be considered beneficial or cost effective. By comparison, a business that is struggling and is looking for ways to improve its efficiency may view a low effect size of .3 as very valuable if the increase in efficiency increases revenues, profits, and investor confidence.

Comparing Means

In comparing means, the index of ES is a ratio that is calculated by dividing the difference between the two means by a standard deviation. (See Chapter 5 for a discussion of standard deviation.) The literature offers several approaches to obtaining the standard deviation that is used as the denominator in the equation to compute the effect size. In this book, we use some of the most commonly used methods that are fairly easy to use and interpret.

In experimental studies the means of experimental and control groups are usually compared in order to assess the effectiveness of the intervention. In the computation of the effect size in experimental studies, the difference in means between the two groups is the numerator. Because the experimental group scores are usually higher than the scores of the control group, the mean of the control group is subtracted from the mean of the experimental group. The standard deviation of the control group is used as the denominator.

$$ES = \frac{Mean_{Exp} - Mean_{Cont}}{SD_{Cont}}$$

Where	ES	$=$	Effect size
	$Mean_{Exp}$	$=$	Mean of experimental group
	$Mean_{Cont}$	$=$	Mean of control group
	SD_{Cont}	$=$	Standard deviation of control group

When the two means in the numerator are two comparison groups, such as Group A and Group B (or men and women), the denominator is the pooled standard deviation of the two groups.

$$ES = \frac{Mean_{GroupA} - Mean_{GroupB}}{SD_{Pooled}}$$

Where	ES	$=$	Effect size
	$Mean_{Group\ A}$	$=$	Mean of Group A
	$Mean_{Group\ B}$	$=$	Mean of Group B
	SD_{Pooled}	$=$	Pooled standard deviation of Group A and Group B

Effect size can also be computed in studies that assess change from pretest to posttest (e.g., in experimental studies). To calculate the effect size in such studies, we first subtract the pretest from the posttest then divide the difference by the standard deviation of the gain (or change) scores.

$$ES = \frac{Mean_{Posttest} - Mean_{Pretest}}{SD_{Gain}}$$

Where	ES	=	Effect size
	Mean $_{Posttest}$	=	Mean of posttest
	Mean $_{Pretest}$	=	Mean of pretest
	SD $_{Gain}$	=	Standard deviation of gain (change) scores

An effect size around .00 indicates that the two groups scored about the same. A positive effect size indicates that the first group listed in the numerator scored higher than the second group or that the posttest mean was higher than the pretest mean. A negative effect size indicates that the second mean listed in the numerator was higher than the first mean.

Studying Relationships

In studying relationships between two variables, effect sizes may be interpreted similarly to the way the correlation coefficient r is evaluated. (See Chapter 8 for a discussion of correlation.) The correlation coefficients serve as an index that quantifies the relationship between two variables and it can be used to evaluate the statistical significance, as well as the practical significance, of the study. Just like d, the index of effect size used to compare means, r can also be positive or negative. However, while the effect size d can take on values that are higher than 1.00, the correlation coefficient r can range only from -1.00 (a perfect negative correlation) to $+1.00$ (a perfect positive correlation). Several researchers suggest the use of squared correlation coefficients (r^2; also known as the *coefficient of determination*) as an index of effect size in place of the correlation coefficient r. Cohen suggests the following guidelines to interpret the *practical* importance of correlation coefficients: $r=.10$ ($r^2=.01$) is considered a small effect; $r=.30$ ($r^2=.09$) is considered medium effect; and $r=.50$ ($r^2=.25$) is considered large effect. Several researchers suggest the use of squared correlation coefficients (r^2; also known as the *coefficient of determination*) as an index of effect size in place of the correlation coefficient r.

Using Samples to Estimate Population Values

In research, we often want to gather data about a population that is of interest to us. However, usually it is not possible or not practical to study all the members of the population. Instead, we select a sample from that population and use the sample's numerical data (the sample statistics) to estimate the population values (the parameters). In using statistics to estimate parameters, we can expect some sampling error.

Population parameters have fixed values but when we select a single sample from a population, the sample statistics are likely to be different from their respective population parameters. Sample values (e.g., the mean) are likely to be higher or lower than the fixed population values they are designed to estimate. Nevertheless, in conducting research we usually use information from a single sample to estimate the parameters of the population from which that sample was selected. Our estimate is expressed in terms of probability, not certainty.

For example, an auditing firm needs to determine if a client company is following Generally Accepted Accounting Principles (GAAP) when recording its accounts receivable. Instead of checking all the individual accounts receivable the company carries, the auditing firm selects a random sample and determines if the company's procedure for accounting for the receivables in the sample is in compliance with GAAP. The evaluation of the sample accounts is then used to estimate how the client company handles all of its accounts receivable.

The auditors may ask themselves how well the results of the examination of the sample receivables represent how the client company records all of its accounts receivables and whether the information gathered from the sample (the sample statistics) are an accurate representation of the population information (the population parameters). Clearly, when we use sample values to estimate population values, instead of studying the whole population, we risk making a certain level of error. All of us can probably agree that a sample is not likely to be a perfect representation of the population. However, in business, as in political polls and other areas of social science, we usually agree to accept a certain "margin of error" and often choose to select a sample, study that sample, and use the information from the sample to make inferences about the population that is of interest to us.

If we select multiple samples of the same size from the same population, compute the sample means, and plot the sample means, we would see that they are normally distributed in a bell-shaped curve. (See Chapter 6 for a discussion of the normal curve.) As the number of samples increases, the shape of the distribution gets closer to a smooth-looking bell-shaped curve.

As an example, suppose we want to calculate the average (i.e., mean) number of miles driven to and from work each day by 50,000 commuters. If we select several samples of 50 commuters each, and ask them how many miles they commute each day, we are not going to get the exact same mean number of miles from each group of 50 commuters. Clearly, there will be some variability in the mean scores obtained from each sample. The expected variation among means that are selected from the same population is considered a *sampling error*. Suppose, further, that we were to select a group of 50 commuters, record their commuting miles, and put their names back in the population list of names, and repeat this process over and over. If we graph the mean number of miles of each sample, we will be able to see that these means would form a normal distribution.

The distribution of the means of multiple same-size samples that are drawn from the same population has its own mean and standard deviation. The mean indicates the location of the center of the distribution. The standard deviation (abbreviated as SD) is an index of the spread of a set of scores and their variability around the mean of the distribution. The SD indicates the average distance of scores away from the mean. (See Chapters 4 and 5, respectively, for a discussion of the mean and standard deviation.) In a normal distribution, approximately 68% of the scores lie within plus or minus 1SD (written as ± 1SD) from the mean; approximately 95% of the scores lie within ± 2SD from the mean; and over 99% of the scores lie within ± 3SD from the mean.

Standard Error of the Mean

The standard deviation of the distribution of multiple sample means that are drawn from the same population is called the **standard error of the mean** and it is expressed by the symbol $SE_{\bar{x}}$ Luckily, we do not have to draw many samples in order to estimate the population mean and standard deviation. Instead, we can draw a single sample and use the information from that sample to estimate the population mean and standard deviation. The mean of the sample is used to estimate the population mean. The standard deviation of the population can be estimated from the sample standard deviation by using this formula:

$$SE_{\bar{x}} = \frac{SD}{\sqrt{n}}$$

Where	$SE_{\bar{x}}$	=	Standard error of the mean
	SD	=	Standard deviation of the sample
	n	=	Sample size

The standard error of the mean is used to estimate the *sampling error*. It shows the extent to which we can expect sample means to vary if we were to draw other samples from the same population. It can be used to estimate the sampling error we can expect if we use the information from a single sample to estimate the population standard deviation.

In the formula used to compute the standard error of the mean, the square root of the sample size is used as the denominator. Therefore, higher sample sizes would result in lower standard errors of the mean. For example, suppose we were to draw two samples from the same population; one sample with 100 members and one sample with 20 members, and compute the standard error of the mean for these two samples. We can expect the standard error of the mean of the sample with 20 members to be higher compared with the larger sample of 100 members. This is because the denominator would be higher when the sample has 100 members compared with a smaller sample of 20. Our estimate of the population standard deviation would be more accurate using the standard error of the mean from the large sample because the standard error of the mean computed from the large sample would be smaller.

The standard error of the mean tells us that if we were to continue to draw additional samples of the same size from the same population, we could expect that 68% of the sample means would be within ± 1SD of our obtained sample mean; 95% of the sample means would be within ± 2SD of the obtained sample mean; and 99% of the sample means would be within ± 3SD of the obtained sample mean.

Confidence Intervals

As was explained, using statistics derived from a single sample to estimate the population parameters is likely to result in a certain level of error. A confidence interval is a way to estimate the population value that is of interest to us. The **confidence interval (CI)** lets us predict, with a certain degree of confidence, where the population parameter is. A confidence interval allows us to state the boundaries of a range within which the population value we try to estimate (e.g., mean) would be included a certain percent (e.g., 95%) of the times in samples of the same size drawn from that same population as our single sample.

The sample mean is used as the center of the confidence interval. The standard error of the means is also used in constructing confidence intervals. The confidence interval includes two boundaries: the lower limit, represented as CI_L; and the upper limit, represented by CI_U.

In many research situations, the goal of the researchers is to use a single sample mean to estimate the population mean. However, there are many other situations where researchers are interested in comparing two means from different populations. An example would be a study where the means of experimental and control groups are compared to each other. (See Chapter 10 for a discussion of the t test for independent samples.) Another example would be a study where researchers may want to study the effect of an intervention by comparing pretest and posttest mean scores for the same group of people. (See Chapter 10 for a discussion of the t test for paired samples.)

When we select multiple samples of the same size from two different populations, or when we repeat a certain intervention with multiple samples of the same size chosen from the same population, the *differences* between the two means are normally distributed. The distribution of the differences between the means has its own mean and standard deviation. The confidence intervals in these studies provide the lower and upper limits of the distribution of the differences between the means.

Computer programs (such as SPSS) usually provide upper and lower limits of the confidence intervals at the 95% confidence level. The interval's lower boundary (CI_L) and its upper boundary (CI_U) are usually reported, along with the mean and standard error of the mean. An interval of 68% confidence (CI_{68}) contains a narrower range compared with a confidence interval associated with 95% confidence. Similarly, intervals of 99% confidence are wider than similar intervals associated with 95% confidence.

The formulas used to construct confidence intervals (that are used to estimate population means in different research situations including two statistics (z scores and t values) that are discussed later (see Chapters 6 and 10). Therefore, no numerical examples are provided here to illustrate how to compute the confidence intervals and their lower and upper limits (CI_L and CI_U).

Steps in the Process of Hypothesis Testing

Research studies that are conducted to test hypotheses follow similar steps. These steps are likely to be taken in studies where samples are selected and studied for the purpose of making inferences about the populations from which they were selected (i.e., inferential statistics).

The hypothesis-testing process starts with the study's research question and ends with conclusions about the findings. Following is a summary of the steps in the process of the statistical hypothesis testing.

1. Formulating a research question for the study.

2. Stating a research hypothesis (i.e., an alternative hypothesis). The hypothesis should represent the researcher's prediction about the outcome of the study and should be testable. Note that a null hypothesis for the study is always implied but it is not formally stated in most cases. The null hypothesis predicts no difference between means or groups, or no relationship between variables.

3. Designing a study to test the research hypothesis. The study's methodology should include plans for selecting one or more samples from the population that is of interest to the researcher; selecting or designing instruments to gather the numerical data; carrying out the study's procedure (and intervention in experimental studies); and determining the statistical test(s) to be used to analyze the data. (See Chapter 15 for further information about the study's methodology.)

4. Conducting the study and collecting numerical data.

5. Analyzing the data and calculating the appropriate test statistics (e.g., Pearson correlation coefficient, t test value, or F ratio; see Chapters 8, 10, and 11, respectively).

6. Determining the appropriate critical value for the obtained test statistic and finding the appropriate p value.

7. Deciding whether to retain or reject the null hypothesis. A p value of .05 is the most commonly used benchmark to decide whether to consider the results statistically significant. If the results are statistically significant, the researcher may also wish to calculate the effect size (ES) to determine the practical significance of the study's results.[8]

8. Making a decision whether to confirm the study's alternative hypothesis (i.e., the research hypothesis) and how probable it is that the results were obtained purely by chance. This decision is based on the decision made regarding the null hypothesis.

9. Summarizing the study's conclusions, addressing the study's research question.[9]

---

8. **A HINT:** The concept of *effect size* is fairly new; therefore, you may come across published research articles that do not include information about the study's effect size.
9. **A HINT:** You may want or need to write a formal report describing the study, which would include a literature review at the beginning of the report and the study's limitations and implications in the discussion section. (See Chapter 15 for guidelines for conducting and reporting research studies.)

And Finally...

Computer programs make the task of data analysis quick, easy, and efficient. However, it is up to the researcher to interpret the results and evaluate their implications. For example, a difference of 3 points between two group means in one study may not be as meaningful as the same finding in another study.

Statistical analyses are based on observations that are collected using certain instruments and procedures. If the instruments used to collect data lack in reliability or validity, any conclusions or generalizations based on the results obtained through using these instruments are going to be questionable. Similarly, when a study is not well designed, one may question the results obtained from such study. Problems resulting from a poorly designed study and bad data cannot be overcome with a fancy statistical analysis. Just because the computer processes the numbers and comes up with "an answer" does not mean that these numbers have any real meaning. Remember what is often said regarding the use of computers: "garbage in-garbage out." This adage applies to the use of statistics as well.

Summary

1. The term **statistics** refers to methods and techniques used for describing, organizing, analyzing, and interpreting numerical data.

2. A **variable** is a measured characteristic that can vary and assume different values or levels.

3. A **constant** is a measured characteristic that has only one value.

4. Variables may be continuous or discrete. **Continuous variables** can take on a wide range of values and contain an infinite number of small increments. **Discrete variables** contain a finite number of distinct values between two given points.

5. **Measurement** is defined as assigning numbers to observations according to certain rules. There are four types of measurements scales: *nominal, ordinal, interval*, and *ratio*.

6. In **nominal scales**, numbers are used to label, classify, or categorize observations to indicate similarities or differences. This is the least precise form of measurement.

7. In **ordinal scales**, observations are ordered to indicate *more than* or *less than* based on magnitude or size. The intervals between the observations, however, cannot be assumed to be equal.

8. In **interval scales**, observations are ordered with equal intervals between points on the scale. Since there is no absolute zero point, inferences cannot be made which involve ratio comparisons.

9. In **ratio scales**, observations are ordered with equal intervals between points. This scale has an absolute zero; therefore, comparisons can be made involving ratios. This is the most precise form of measurement. Ratio scales are generally used in physical sciences rather than in the behavioral sciences.

10. A **population** is the entire group of persons or elements that have some characteristic in common.

11. A **sample** is a group of observations selected from the total population.

12. Some chance variation in sample numerical values (e.g., mean) occurs when we repeatedly select same-size samples from the same population and compare their numerical values. This error, called a **sampling error**, is beyond the control of the researcher.

13. A **parameter** is a measure of a characteristic of the entire population.

14. A **statistic** is a measure of a characteristic of the sample.

15. A sample should be **representative** of the population, because the statistics gained from the sample are used to estimate the population parameters.

16. Most research studies in business are designed to study populations by using samples that are representative of these populations.

17. In selecting a **simple random sample**, every member of the population has an equal and independent chance of being selected. The table of random numbers is often used to draw a random sample.

18. In **systematic sampling**, every *Kth* person is selected from the population. *K* is determined by dividing the total number of population members by the desired sample size.

19. The first step in obtaining a **stratified sample** is to divide the population into subgroups (called *strata*), followed by a random selection of members from each subgroup. The final sample represents, proportionately, the various subgroups in the population.

20. A **convenience** (or **incidental**) **sample** is a sample that is readily available to the researcher. Researchers have to exercise great caution in generalizing results from a convenience sample to the population.

21. **Sample bias** refers to *systematic* (as opposed to *random*) differences between the sample and the population from which it was selected. A biased sample contains a certain systematic error.

22. As sample size increases, it is more likely to be an accurate representation of the population, especially when the sample is randomly chosen. In many research studies, a sample size of at least 30 is desirable. However, size alone does not guarantee that the sample is representative and a large sample may still be biased.

23. There is a clear relationship between the sample size and the confidence level in rejecting the study's null hypothesis. As the sample size *increases*, a *lower* computed test statistic value is needed in order to reject the null hypothesis at the $p=.05$ level.

24. When the population is greater than 10,000, a sample size of 1,000 - 1,500 (10%-15%) may accurately represent this population.

25. **Parametric statistics** (also called **assumption-free statistics**) are applied to populations that meet certain requirements. **Nonparametric statistics** can be applied to all populations, even those that do not meet the basic assumptions.

26. Parametric statistics are used more often by researchers and are considered more powerful and more efficient that nonparametric statistics.

27. Nonparametric statistics can be used with nominal, ordinal, interval, and ratio scales. Parametric statistics can be used with interval and ratio scales only.

28. **Descriptive statistics** classify, organize, and summarize numerical data about a particular group of observations.

29. **Inferential statistics** (also called **sampling statistics**) involve selecting a sample from a defined population, studying the sample, and using the information gained to make inferences and generalizations about that population.

30. Descriptive and inferential statistics are not mutually exclusive and the same measures can be used in both types. The purpose or the use of the statistics determines whether they are descriptive or inferential.

31. A **hypothesis** is a prediction (an "educated guess") about the outcome of the study. After the hypothesis is proposed, a study is designed to test the hypothesis.

32. The main hypothesis proposed by the researcher about the study's outcome is called the **alternative hypothesis** (or simply the **hypothesis**). It is represented by the symbol H_A or H_1.

33. A **null hypothesis** (also called a **statistical hypothesis**) always states that there would be no differences between groups or means being compared, or no relationship between variables being correlated, beyond what might be expected purely by chance.

34. Hypotheses may be stated as directional or nondirectional. A **directional hypothesis** predicts the direction of the outcome of the study. A **nondirectional hypothesis** predicts that there would be a statistically significant difference or relationship, but the direction is not stated.

35. In studies where *differences* are investigated, a directional hypothesis predicts which group would score higher on the dependent variable. A nondirectional hypothesis predicts a difference in scores on the dependent variable, but not the direction of the difference (i.e., which group's mean would be higher) and a null hypothesis predicts no difference between the means.

36. In studies of *relationship*, a directional hypothesis predicts whether the relationship (e.g., correlation) would be positive or negative. A nondirectional hypothesis predicts that the variables would be related, but it does not specify whether the relationship would be positive or negative. A null hypothesis predicts no relationship between the variables.

37. The process of the statistical hypothesis testing starts with a decision regarding the null hypothesis.

38. The study's statistical results are used to decide whether the null hypothesis should be retained or rejected. In studies where the alternative hypothesis is directional or nondirectional, **rejecting** the null hypothesis usually leads to the confirmation of the alternative hypothesis, while **retaining** the null hypothesis usually leads to a decision not to confirm the alternative hypothesis.

39. Results may be reported as **statistically significant** or **not statistically significant**. When the results are statistically significant, the *exact* level of significance may be reported.

40. Statistical results are reported in terms of probability, not certainty. Results that are statistically significant are usually reported in terms of **probability (p value)**, or **level of significance**, using terms such as $p<.05$ or $p<.01$.

41. When the probability level is set at the beginning of the study, *before* collecting and analyzing the data, it is represented by the Greek letter **alpha (α)**. The convention is to use an alpha level of .05.

42. A **Type I error** is made when researchers decide to *reject* the null hypothesis (H_0) when in fact it is true and *should not be rejected*.

43. A **Type II error** is made when researchers decide to *retain* the null hypothesis, when in fact *it should be rejected*.

44. The proper decision is made when researchers *reject a false* null hypothesis, or when they *retain a true* null hypothesis.

45. Tables of critical values are used to determine the appropriate p value (level of statistical significance). To consult these tables, the appropriate **degrees of freedom (df)** need to be calculated. In most statistical tests, df is calculated as n-1 (the number of people in the study, minus 1). Researchers routinely get both df and p values from computer statistical programs.

46. In interpreting statistical results, researchers should look at the *direction* of the outcome and whether the results are *statistically significant*. When appropriate, *effect size* should be computed to evaluate the *practical significance* of the data.

47. **Effect size (ES)** is an index that is used to express the strength or magnitude of difference between two means or the relationship between variables. Using the index of effect size allows researchers to evaluate the *practical significance*, in addition to the *statistical significance* of their studies.

48. There are several ways to calculate effect sizes. One of the most commonly used effect size, called *d*, was developed by Cohen (1988). ES is calculated by dividing the difference between the two means by a standard deviation (SD).

$$ES = \frac{Mean_1 - Mean_2}{SD}$$

49. While no clear-cut guidelines are available for the interpretation of effect size, many researchers follow those offered by Cohen (1988): *ES* of .2 is considered small; *ES* or .5 is considered medium effect; and *ES* of .8 is considered large. Guidelines developed by other researchers define effect size of .2 as small; effect size of .6 is moderate; effect size of 1.2 is large; and effect size of 2.0 is very large. Effect sizes of 1.00 or higher are rare in business and other types of social science research.

50. In studying relationships between variables, the effect size may be interpreted similarly to the way the correlation coefficient r is evaluated. Researchers use r or r^2 as an index of effect size. Cohen suggested the following guidelines for evaluating correlation coefficients: $r=.10$ ($r^2=.01$) is considered small; $r=.30$ ($r^2=.09$) is considered medium; and $r=.50$ ($r^2=.25$) is considered large.

51. Population parameters have fixed values but when we select a single sample from a population, the sample statistics are likely to be different from the respective population parameters. Sample values (e.g., means) are likely to be higher or lower than the fixed population values they are designed to estimate.

52. If we select multiple samples of the same size from the same population, compute the sample means, and plot the sample means, we would see that they are normally distributed in a bell-shaped curve.

53. The distribution of multiple sample means of the same size that are drawn from the same population has its own mean and standard deviation.

54. The standard deviation of the sample means is called the **standard error of the mean.** A single sample can be used to compute the standard error of the mean, which can be used as an estimate of the standard deviation of the population. The sample standard deviation and size are used in the formula to compute the standard error of the mean:

$$SE_{\overline{x}} = \frac{SD}{\sqrt{n}}$$

55. The index of the standard error of the mean is used to estimate sampling error. It shows the extent to which we can expect sample means to vary if we were to draw other samples from the same population. It can be used to estimate the sampling error we can expect if we use the information from a single sample to estimate the population standard deviation.

56. The standard error of the mean tells us that if we were to continue to draw additional samples of the same size from the same population, we could expect that 68% of the sample means would be within ± 1SD of our obtained sample mean; 95% of the sample means would be within ± 2SD of the obtained sample mean; and 99% of the sample means would be within ± 3SD of the obtained sample mean.

57. The **confidence interval (CI)** allows us to state the boundaries of a range within which the population value (e.g., the mean) we try to estimate would be included. The interval lets us predict, with a certain degree of confidence, where the population parameter is expected to be.

58. The sample mean serves as the center of the interval used to estimate the population mean. The standard error of the means is also used in constructing confidence intervals.

59. The confidence interval includes two boundaries: the lower limit, represented as $\mathbf{CI_L}$; and the upper limit, represented by $\mathbf{CI_U}$.

60. When we select multiple samples of the same size from two different populations, or when we repeat a certain intervention with multiple samples of the same size chosen from the same population, the *differences* between the two means are normally distributed. This distribution of differences between the means has its own mean and standard deviation. The confidence intervals in these studies provide the lower and upper limits of the distribution of the differences between the means.

61. The confidence level of 95% (CI_{95}) is used the most and it is the one that is reported most often on the printouts of statistical software programs (e.g., SPSS). The interval's lower boundary (CI_L) and its upper boundary (CI_U) are usually reported, along with the mean and standard error of the means.

62. Research studies that are conducted to test hypotheses follow similar steps. The hypothesis-testing process starts with the study's research question and ends with conclusions about the findings. These steps are likely to be taken in studies where samples are studied for the purpose of making inferences about the populations from which they were selected.

Part Two

Descriptive Statistics

Chapter 3

Organizing and Graphing Data

Organizing Data

Frequency Distributions

Organizing and graphing data allows researchers to describe, summarize, and report their data. By organizing data, they can compare distributions and observe patterns. In most cases though, the original data we collect is not ordered or summarized. Therefore, after collecting data, we may want to create a **frequency distribution** by ordering and tallying the scores.

To illustrate the use of frequency distributions, let's look at the following example. A finance professor wants to assign end-of-term letter grades to the 25 students in her Principles of Finance class. After administering a 30-item final examination, the professor records the students' test scores next to each students name. Table 3.1 shows the number of correct answers obtained by each student on the Principles of Finance final examination. Next, the professor can create a frequency distribution by ordering and tallying these test scores (Table 3.2).

Table 3.1   **Scores of 25 Students on a 30-Item Test**

27	16	23	22	21
25	28	26	20	22
30	24	29	17	28
24	17	23	24	26
19	21	18	23	25

Table 3.2   **A Frequency Distribution of a 30-Item Test Ordered from the Highest to the Lowest Score: Test Scores of 25 Students**

Score	Frequency	Score	Frequency
30	1	22	2
29	1	21	2
28	2	20	1
27	1	19	1
26	2	18	1
25	2	17	2
24	3	16	1
23	3		

Note that the list of scores is still quite long; the highest is 30 and the lowest is 16. The professor may want to group every few scores together in order to assign letter grades to the students. Following is a discussion of the process used for grouping scores.

Class Intervals

When the ordered list of scores in the frequency distribution is still quite long, as is the case in our example (see Table 3.2), the professor may want to group every few scores together into **class intervals**. Class intervals are usually created when the range of scores is at least 20. The recommended number of intervals should be 8-20.[1] The biggest disadvantage of using class intervals is that we lose details and precision. That is, because scores are grouped, we cannot tell what the exact score obtained by each person was. For example, assume we know that there are 4 students in a class interval of 20-25. We cannot tell which score was obtained by each of these students, only that their scores were between 20 and 25.

Two rules should be observed when creating class intervals: (a) all class intervals should have the same width; and (b) all intervals should be mutually exclusive (i.e., a score may not appear in more than one interval). Whenever possible, the width of the interval should be an odd number to allow the midpoints of the intervals to be whole numbers.

Table 3.3 contains test scores of 30 students on an 80-point test. The lowest score obtained by a student on the test is 31 and the highest score is 80. The scores in the table are listed in descending order. Table 3.4 shows the same scores grouped into 10 class intervals. Each interval has a width of 5 points.

Table 3.3   **Test Scores of 30 Students**

80	62	57	52	44
74	61	57	51	43
69	59	56	50	41
66	58	55	49	39
65	58	54	48	36
63	57	53	47	31

---

1.   **A HINT:** There are no strict rules as to the number of class intervals that should be used. Some textbooks recommend 8-20 intervals while others recommend 10-20. In the examples in this chapter we used 10 intervals (Table 3.4) and 8 intervals (Table 3.8). Most computer statistical programs are likely to create class intervals for you, so the computational steps in the book are used mostly to explain the concept of *class intervals*.

Table 3.4	A Frequency Distribution of 30 Scores With Class Intervals of 5 Points and Interval Midpoints	

Class Interval (5 points)	Midpoint	Frequency
76-80	78	1
71-75	73	1
66-70	68	2
61-65	63	4
56-60	58	7
51-55	53	5
46-50	48	4
41-45	43	3
36-40	38	2
31-36	33	1

Cumulative Frequency Distributions

Another way to organize data is to create a *cumulative frequency distribution*. **Cumulative frequencies** show the number of scores *at* or *below* a given score. Percentages are often added to these tables. Table 3.5 is a cumulative frequency table showing test scores of 20 students.

Table 3.5 starts with a frequency distribution in column 1 and column 2 (similar to the frequency distribution in Table 3.2). These two columns are titled *Score* and *Frequency*.

To create the third column, *Percent Frequency*, we convert into percentages the frequencies that are listed in the second column. For example, inspecting the top of column 2, we can see that 1 student had a score of 20. Because there are 20 students in the class, we can say that 5% (1 out of 20) of those students had a score of 20. Similarly, we can see that 10% (2 students out of 20) had a score of 17.

The next step is to add the fourth column, titled *Cumulative Frequency*. To construct this column, you have to work your way from the bottom up. To calculate the first entry at the bottom of the fourth column, look at the lowest numbers in columns 1 and 2 and ask yourself the following: How many students had a *score of 5* (the lowest score in this distribution) or *less*? The answer is "1;" therefore, "1" is recorded at the bottom of column 4.

Next, ask yourself: How many students had a *score of 6 or less*? (To find the answer to this question, add up the two lowest numbers in column 2) The answer is "3" and this number is recorded right above "1" at the bottom of the fourth column. Now, calculate the number of students who had a *score of 8* or *less*. The answer is "5," the third lowest number in column 4. Continue constructing column 4 all the way up until the column is completed.

Table 3.5   Cumulative Frequency Distributions of Test Scores of 20 Students [2]

(Col. 1) Score	(Col. 2) Frequency	(Col. 3) Percent Frequency	(Col. 4) Cumulative Frequency	(Col. 5) Cumulative Percentage
20	1	5	20	100
19	1	5	19	95
17	2	10	18	90
16	4	20	16	80
14	4	20	12	60
10	3	15	8	40
8	2	10	5	25
6	2	10	3	15
5	1	5	1	5
	$N=20$			

The fifth column is titled *Cumulative Percentage*. To create this column, the cumulative frequencies in column 4 are converted to percentages. The conversion can be done by starting either at the top or at the bottom of column 5. To compute each entry in column 5, convert the corresponding cumulative frequency in column 4 into percentages. To do so, divide the number in column 4 by 20, the total number of students in our example. If you work your way from the top down, the first cumulative frequency you need to convert to percentage is 20 (the top number in column 4). In other words, you have to calculate what percentage of students in the class had a score of 20 or less. Clearly, *all* 20 students in the class had a score of 20 or less. Therefore, enter 100 (i.e., 100%) at the *top* of the fifth column. Next, convert 19 (the second number in column 4) into percentages. Nineteen out of 20 (the total number of students in the class) is 95%, which appears as the second number in column 5. The third cumulative frequency in column 4 is 18, which converts to 90% (because 18 out of 20 is 90%). Continue to work all the way down to the bottom of column 5 until it is complete.

Using the same steps as those used to create the entries in Table 3.5, professors can use scores from tests they give their students to compute the students' cumulative percentiles. These cumulative percentiles are also called *percentile ranks* and they are used to compare performance of students in the class to their classmates (See Chapter 6 for a more comprehensive discussion of percentile ranks). Using the data in Table 3.5 as an example, we can say that a student with a score of 19 had a percentile rank of 95. This percentile rank means that the student did better than, or as well as, 95% of the other students in the class who took the examination at the same time as that student. Similarly, a student with a score of 14 had a percentile rank of 60 and did better than, or as well as, 60% of the students in the class.

---

2.   **A HINT:** Note that the number of scores in this distribution is indicated at the bottom of the second column as "N=20." In statistics, the symbol *N* or *n* is used to represent the number of cases or scores.

Graphing Data

Graphs are used to communicate information by transforming numerical data into a visual form. Graphs allow us to see relationships not easily apparent by looking at the numerical data. There are various forms of graphs, each one appropriate for a different type of data. While many computer software programs provide a dazzling array of graphic choices, it is the responsibility of those creating the graphs to select the right graph for their data. The rest of this chapter discusses various graphs and how they can be used.

Histogram and Frequency Polygon

Frequency distributions, such as the one in Table 3.6, can be depicted using two types of graphs, a **histogram** (Figure 3.1, *Part a*) or a **frequency polygon** (Figure 3.1, *Part b*).

Table 3.6   **A Frequency Distribution of 13 Scores**

Score	Frequency
6	1
5	2
4	4
3	3
2	2
1	1

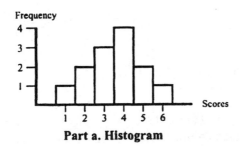

Part a. Histogram

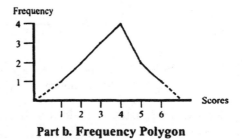

Part b. Frequency Polygon

Figure 3.1   Histogram and frequency polygon of the data in Table 3.6

In drawing histograms and frequency polygons, the vertical axis *always* represents frequencies, and the horizontal axis *always* represents scores or class intervals. The lower values of both vertical and horizontal axes are recorded at the intersection of the axes (at the bottom left side). The values on both axes increase as they move farther away from the intersection (Figure 3.2).

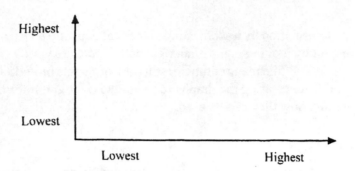

Figure 3.2   Graphing the lowest and highest values on each axis

When possible, the frequency axis (the vertical axis) of histograms and frequency polygons should start with a frequency of 0 (zero). The scores axis (the horizontal axis) may start with a score of 0, *or* with a higher score. If the first score to be marked on the horizontal axis is *higher than 1*, we can indicate the skipping of lower scores by showing a break in the axis. For example, if the lowest score to be graphed is 15, we indicate skipping the first 14 scores as shown in Figure 3.3.

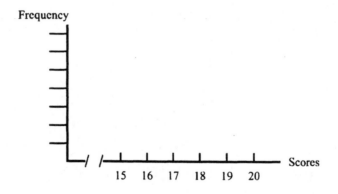

Figure 3.3   Graphing frequency distributions when the lowest score
on the horizontal axis is higher than 1

When scores (or intervals) in a distribution have a frequency of 0 (zero) (see Table 3.7), histograms or frequency polygons created to depict that distribution should be drawn as demonstrated in Figure 3.4. (If you use a software package to draw graphs, you may want to check the options available on that package).

Table 3.7   **A Frequency Distribution of 11 Scores**

Score	Frequency
6	1
5	2
4	4
3	3
2	0
1	1

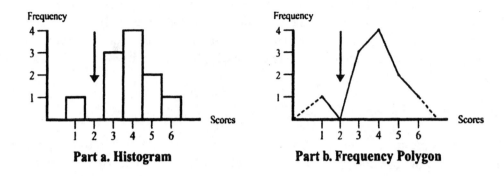

Part a. Histogram   **Part b. Frequency Polygon**

Figure 3.4   Histogram and frequency polygon of the data in Table 3.7

Both the histogram and the frequency polygon can be used to graph individual scores, as well as scores grouped into class intervals. When graphing class intervals, tick marks on the horizontal axis show the interval midpoint. The upper and lower scores in each interval may also be recorded instead of the interval midpoint. Table 3.8 shows Statistics test scores of 25 students, and Figure 3.5 presents the same data using a frequency polygon.

Table 3.8   A Frequency Distribution of 25 Scores With Class Intervals and Midpoints

Class Interval	Midpoint	Frequency
38-42	40	1
33-37	35	3
28-32	30	4
23-27	25	6
18-22	20	5
13-17	15	3
8-12	10	2
3-7	5	1

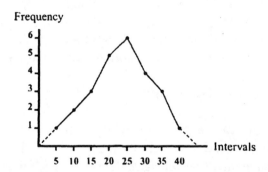

Figure 3.5   A frequency distribution with interval midpoints representing class intervals of the data in Table 3.8

As the number of scores in the distribution increases, it is more likely to have smoother curve lines, compared to a distribution with fewer scores. Distributions with a small number of cases are likely to have curve lines that are more jagged and uneven. The bell-shaped normal distribution that all of us are familiar with is actually a special case of a frequency polygon with a large number of cases. (See Chapter 6 for a discussion of the normal curve and bell-shaped distributions).

Comparing Histograms and Frequency Polygons

To some extent, the decision about whether to use a histogram or a frequency polygon is a question of personal choice. One advantage of the polygon over the histogram is that the polygon can be used to compare two groups by displaying both groups on the same graph. For example, Figure 3.6 shows test scores of a group of men and a group of women on a management test. As can be easily seen, the ranges of both groups were about the same. However, the women performed overall better than the men and obtained higher scores.

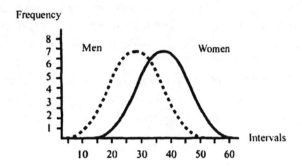

Figure 3.6   A frequency polygon comparing scores of men and women

Several textbooks list another difference between a histogram and a frequency polygon. These books suggest using histograms for data that naturally occur in discrete units. For example, the number of employees who work for a company may be reported to be 101, 102, or 103, but not 101.5 or 102.5.[3] By comparison, frequency polygons are recommended for depicting distributions of scores that can take on in-between values, such as the weight or size of a product. For example, let's say we want to use a frequency polygon to graph the miles per gallon of gas used by a company's trucks, rounded off to the nearest mile per gallon. Using 1-mile per gallon increments to mark the horizontal axis, the frequency polygon would show only data points such as 11, 12, or 13 miles per gallon even though in reality, trucks may drive a number of miles per gallon that is somewhere *between* 11 and 12, or *between* 12 and 13. It is important to remember that although the scores on the variable being graphed are measured on a continuum, the frequency polygon itself may show only discrete units. In our example, we use a frequency polygon with 1-mile per gallon increments, rather than record the exact miles per gallon for each truck.

Graphing Cumulative Frequency Distributions

Cumulative frequency distributions can be depicted visually using a graph called an **ogive** (pronounced "oh-jive"). The horizontal axis on the ogive is used for recording the scores and the vertical axis is used for recording the cumulative frequencies or the cumulative percentages. (See Table 3.5, columns 4 and 5) A typical ogive tends to be steep in the middle and flat at the lowest and highest points, especially when the distribution includes many scores. This is why the ogive is also called an **"s" curve.**

---

3.   **A HINT:** Although the number of employees are reported only as whole numbers, *group* mean score may have decimal places, such as 103.4 or 104.8

Looking at Figure 3.7, we can read the same information as that presented in Table 3.5. For example, we can determine that a person with a score of 17 scored as well as, or better than, 90% of those who took the test.

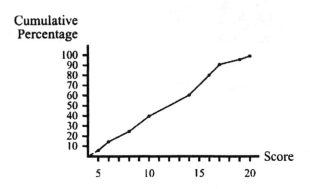

Figure 3.7   An ogive of a cumulative distribution showing the data in Table 3.5

Pie Graph

The **pie graph** (or **pie chart**) looks like a circle that is divided into "wedges," or "segments." Each wedge represents a category or subgroup within that distribution. The size of each wedge indicates the percent of cases represented by that wedge. By inspecting the pie graph, we can readily see the proportion of each wedge to the total pie, as well as the relationship among the different wedges. The percentages represented by the different-sized wedge should add up to 100%.

When drawing a pie graph, the different wedges of the pie should be identified, and numerical information, such as percentages, should be included. This would allow easy and accurate interpretation of the graph. There should not be too many wedges in the pie circle. The *Publication Manual* of the American Psychological Association (APA) (2001, p. 179) recommends that the pie graph be used to compare no more than 5 items.[4] However, it is not uncommon to see reports that include pie graphs with 6-7 wedges. APA *Publication Manual* also recommends that the wedges be ordered from the largest to the smallest, starting at 12 o'clock.

To illustrate how to draw and interpret pie graphs, study the data in Table 3.9 and the graph in Figure 3.8. The table and pie graph show the proportions of consumers' preferences for toppings on pizza. Note that the total percentages of the toppings add up to 100%.

---

4.   **A HINT:** The complete reference for the APA publication manual is: American Psychological Association. (2001). *Publication manual of the American Psychological Association* (5ᵗʰ ed.). Washington, DC: Author.

Table 3.9   Proportion of Respondents Who Prefer Different Pizza Toppings

Topping	Percent
Mushrooms	20
Pepperoni	35
Vegetables	35
Anchovies	10
TOTAL	100

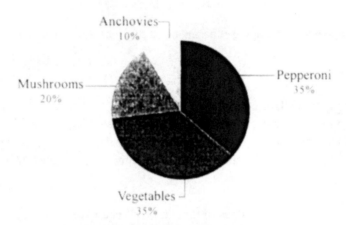

Figure 3.8   A pie graph showing the percentages of students that prefer each pizza topping

Pie graphs lend themselves well to making comparisons by placing two pie graphs next to each other. For example, we can create two side-by-side pies to show changes in demographic characteristics (e.g., racial or ethnic groups) in a city. One pie graph can depict the demographic data from one year and the other graph can show the same demographic data from another year. This would allow those studying the two graphs to see changes from one year to the next. Or the two pies can represent demographic data from two different cities to allow a comparison of the two. Another popular way to use two pies side by side is to use one of the pies for income and one pie for expenses. To show that the budget is balanced, the total dollar amount in the pie showing expenses should not exceed the dollar amount in the pie showing the income. (For example, the pies can show income and expenses of a business) By looking at the two pies, it is easy to see what the main sources of income are and what the main expenses are.

Bar Graph

A **bar graph** (also called a **bar diagram**) is a graph with a series of bars next to each other. The bars often represent *discrete* categories and they are ordered in some way, usually from the highest to the lowest, or from the lowest to the highest. The bars are placed next to one another but they *should not touch* each other.

To illustrate the use of a bar graph, let's say we want to examine the support for ethics training among four types or organizations: manufacturers, wholesalers, retailers, and non-profits. Five hundred respondents from each group indicated whether they require their employees to attend ethics training and the results are recorded in Table 3.10. The actual number of respondents in each type of organization that require ethics training was converted into percentages and recorded in the right-hand column of the table. Inspecting the data in Table 3.10 makes it clear that there are differences among manufacturers, non-profits, retailers, and wholesalers regarding their support of ethics training.

Table 3.10   Comparing the Responses of Manufacturers, Non-Profits, Retailers and Wholesalers Organizations to the Question: "Do You Require Employees to Attend Ethical Training?"

Type of Organization	% Requiring Ethics Training
Manufacturers	85
Non-Profits	68
Retailers	30
Wholesalers	77

Note that the four types of organizations in Table 3.10 comprise categories that are independent of each other. In other words, the responses of people in one type of organization do not affect or change the responses of people in another. Note also that the category percentages in the column on the right side do not add up to 100% because each row represents an independent category (a group of respondents). This is different from pie graphs where the percentages add up to 100%. For example, if we look again at Table 3.9, that is used to draw the pie graph in Figure 3.8, we can see that the percentages in the right-hand column add up to 100%.

Before drawing the bar graph in Figure 3.9, we need to decide how to order the bars. Although the types of organizations are listed in alphabetical order in the "Type of Organization" column (Table 3.10) we need to reorder the groups so that the bar graph we draw has bars that are ordered by height. Therefore, we can start with the bar representing the manufacturers, the group with the highest percentage (85%). This bar is followed by the bars representing the other three groups: Wholesalers (77%), Non-Profits (68%), and Retailers (30%).

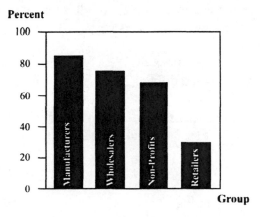

Figure 3.9   A bar graph showing the data in Table 3.10

The bars may also be drawn horizontally, to allow for easier identification and labeling of each bar. Figure 3.10 shows the same data as in Figure 3.9

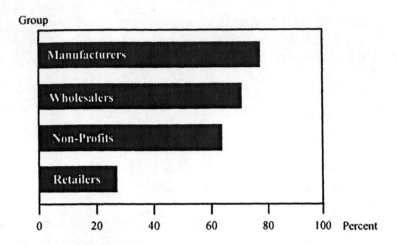

Figure 3.10   A horizontal bar graph showing the same data in Figure 3.9

Occasionally, we may want to compare data from two or more groups *within* each category, in addition to comparing the data *across* categories. To facilitate such comparison, we can use **joint bars**.[5] Unlike the bars representing different categories in bar graphs (e.g., types or organizations) the joint bars *do touch* each other.

Let's say we want to compare the ratio of female and male students who are enrolled in different college majors programs: Accounting, Finance, Management, and Marketing. Table 3.11 displays gender data across the 4 majors. After inspecting the table, we can see that there are differences in the proportions of female and male students enrolled in each of the 4 college majors. An overwhelming majority of students in Finance are female (85%) and the majority of students in Management are male (76%). While there are clear differences between the genders in Finance and Management, these differences are not as great as in Accounting and Marketing.

The joint bars in Figure 3.11 represent the data in Table 3.11. The differences between the lengths of the joint bars in each of the four majors express visually the differences in gender enrollment in these majors.

Table 3.11   Enrollment Breakdown by Gender in Four Undergraduate College Majors (in percents)

Major	Females	Males	Total
Accounting	41	59	100
Finance	85	15	100
Management	24	76	100
Marketing	57	43	100

5.   **A HINT:** You should not use too many joint bars in each category in bar graphs. Having too many bars creates a graph that is difficult to follow and interpret. Most researchers do not use more than 6 joint bars and in most published research you are likely to see 2-5 joint bars.

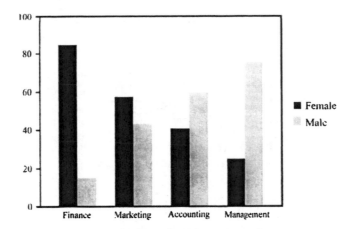

Figure 3.11 A bar graph with joint bars comparing enrollment figures of male and female students in four different college majors

When drawing a bar graph with joint bars, it may be difficult at times to decide how to order the bars. For example, if we had decided to order the bars by male student enrollment figures, then the joint bars representing *Management* should be first and the joint bars representing *Finance* enrollment should be last. In Figure 3.11, the joint bars are ordered by female student enrollment. Therefore, the first set of joint bars shows the gender enrollment figures representing *Finance* and the last set of joint bar shows the enrollment figures representing *Management*.

Bar graphs may look a bit like histograms because both of them have a series of bars next to each other. However, they are used for different purposes. The histogram, as you remember, is used to display frequency distributions. The horizontal bar shows scores or class intervals and the vertical axis shows frequencies. The scores (or intervals) in a histogram comprise an interval or a ratio scale, and they are on a continuum in numerical order. By comparison, the bars in a bar graph represent nominal, categorical data (e.g., majors in college or names of companies), and *do not* imply a continuum. To indicate that each bar is independent of the other bars and represent discrete data, these bars should not touch.

Line Graph

A **line graph** is used to show relationships between two variables, which are depicted on the two axes. The *horizontal* axis indicates values that are on a continuum (e.g., calendar years or months). The *vertical* axis can be used for various types of data (e.g., test scores, sales, income). A line connects the data points on the graphs. Table 3.12 shows mean salaries of salespersons over four years, and Figure 3.12 shows the same information graphically.

Table 3.12   **Mean Salaries of Salespeople, 2003-2006**

Year	Mean Salary (in thousands)
2003	56
2004	74
2005	66
2006	81

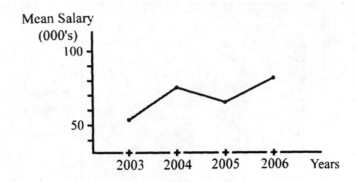

Figure 3.12   A line graph showing the data in Table 3.12

A big advantage of the line graph is that more than one group can be shown on the same graph simultaneously. If you cannot use colors, each group can be presented by a different kind of line (e.g., broken or solid). Figure 3.13 shows mean salaries of two groups over a four-year period.

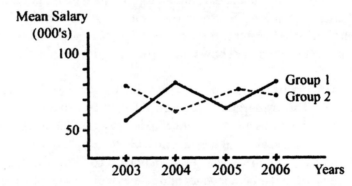

Figure 3.13   A line graph showing mean salaries of two groups over a four-year period

Notice that the line graph is different from a frequency polygon. Although the two graphs may look somewhat similar at times, the line graph is used for a different purpose and it does not display frequencies on the vertical axis.

Box Plot

The **box plot**, which is also called **box-and-whiskers**, was developed by John Tukey in 1977. The graph consists of a box and two whiskers and is used to depict the spread and skewness of frequency distributions. The box represents the middle 50% of the distribution and the whiskers represent the top and bottom 25% of the distribution. A horizontal line inside the box represents the median of the distribution.[6] The lengths of the whiskers indicate the spread of the distribution.

To create the box plot, the scores are first ordered and divided into four quartiles, identified as Q_1, Q_2, Q_3, and Q_4. The two middle quartiles (Q_2 and Q_3) are located *within* the box, whereas the two extreme quartiles (Q_1 and Q_4) are displayed using vertical lines (the whiskers) *outside* the box.

---

6. **A HINT:** See Chapter 4 for a discussion of the median.

To illustrate how to use, construct, and interpret a box plot, let's look at the following example. A company's quality control manager implements a new quality system to increase the number of batches of product that meet the company's quality standards. From a sample of 20 batches, the quality control manager recorded the number of batches that met the company's quality standard before implementing the new system. This is recorded in the "Old System" column of Table 3.13. After implementing the new quality system for one month, a sample of 20 more batches produced the number of acceptable batches that are recorded in the "New System" column of the table. Following are the Old System and New System scores for the 20 batches from each sample (Table 3.13).

To display the changes in quality, the quality control manager creates a box plot graph. Figure 3.14 includes two box plots, one for the Old System scores and one for the New System scores. Notice that the median scores of the Old System and New System are similar (31.0 and 30.5, respectively). However, the range of the scores on the Old System is much higher than the range of the New System. The scores on the Old System range from 13-49 compared with a range of 24-36 on the New System. Note that the whiskers of the Old System scores are much longer indicating a higher range of scores in the top and bottom quartiles, compared with the Old System whiskers that are quite short. Further, the two middle quartiles (Q_2 and Q_3) that include the middle 50% of the scores (the "boxes") are narrower on the New System compared with the Old System.

Table 3.13	Acceptable Number of Products: Old System and New System	
Batch Number	Old System	New System
1	13	24
2	18	25
3	22	25
4	24	26
5	27	27
6	28	28
7	28	28
8	29	29
9	29	29
10	30	30
11	32	31
12	32	31
13	33	32
14	35	32
15	38	32
16	39	33
17	39	34
18	42	34
19	45	35
20	49	36

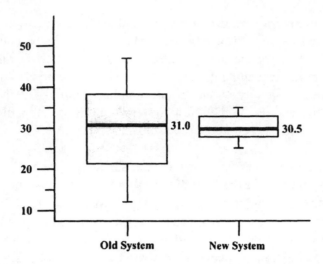

Figure 3.14　A box plot of the Old System and New System scores showing the data in Table 3.13

Based on the data in Table 3.13 and the box plot in Figure 3.14, the quality control manager observes the following: (a) The most extreme Old System scores were eliminated, and (b) the range of scores decreased from Old System to New System, as can be seen by the shorter whiskers on the New System. The quality control manager concludes that there is improved consistency of the products being manufactured even though the median remained approximately the same.

Drawing Accurate Graphs

This chapter describes some of the most popular and commonly used graphs, as well as some of the rules and guidelines for drawing such graphs. Although you may not be able to always follow these guidelines because most of us depend on statistical software packages to generate graphs for us, you should carefully check the different options available on the software packages that you use. It is the responsibility of those creating the graphs to choose the right graphs for their data and to create graphs that are accurate and clearly labeled.

Graphs should offer an accurate visual image of the data they represent and be easy to read and interpret. Each graph should be clearly labeled and different parts of the graph should be identified. Graphs that are too "dense" or include too much information can be confusing and hard to follow. This is especially true for graphs in printed materials where colors are not available (e.g., hard copies of journals). For example, pie graphs with too many wedges or line graphs that depict too many groups may become too "dense" and difficult to read.

To illustrate this point, look at the line graph in Figure 3.15. This graph shows mean incomes (in thousands of dollars) of 6 groups of employees over 4 years. Having so many lines in one graph makes it difficult to compare the groups or observe trends over time.

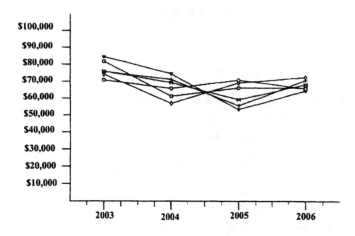

Figure 3.15   Line graph showing too many horizontal lines (6 lines) that create
a confusing graph

The *scale* one chooses for drawing graphs can affect the message conveyed by the graph. To illustrate this point, let's look at Figure 3.16 and study the two graphs in *Part a* and *Part b*. This figure shows the job placement rates (in percentages) of MBAs from four graduate schools. Note that while the two graphs look similar, the vertical axis in *Part a* starts with a score of 0 (zero) while the vertical axis in *Part b* starts with a score of 50. Consequently, the differences between the four bars that represent the graduate schools in *Part b* look much more pronounced than in *Part a*. While the scale of the bars in *Part b* shows more details, it also magnifies the differences and presents an exaggerated picture of the differences.

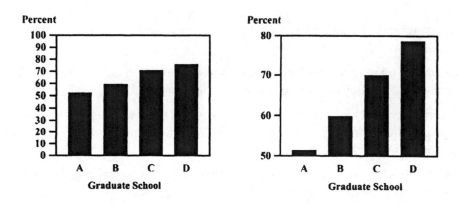

Figure 3.16 Placements of four MBA programs: A comparison of two graphs showing
the same data

Summary

1. The first step in organizing data is to *order* the scores in some manner. For example, the scores can be ordered from the highest to the lowest score or from the lowest to the highest score.

2. When scores in a distribution repeat more than one time, researchers may choose to *tally* the scores in order to eliminate the need to list duplicate scores.

3. A **frequency distribution** is a list of scores that have been ordered and tallied.

4. When a distribution has a wide range of scores (a range of 20 or more in most cases), it is recommended that every few scores in that distribution be grouped into **class intervals**. However, when class intervals are used, detail and precision are lost and there is no way to know what the *exact* scores in each interval are.

5. When using class intervals, all intervals should have the same width and no score should appear in more than one interval. Using intervals of odd width is recommended so that the midpoints of the intervals are whole numbers.

6. Data may be organized using a **cumulative frequency distribution**, which shows the number and percentage of scores *at* or *below* a given score.

7. **Graphs** convey numerical information in a visual form. It is important to choose the proper graph for each type of data.

8. **Histograms** and **frequency polygons** are used to visually display frequency distributions.

9. When drawing either a histogram or frequency polygon, the following guidelines should be used: (a) The vertical axis represents frequencies and the horizontal axis represents scores or class intervals, and (b) the lower values of both axes are at the point of their intersection. (Other guidelines for drawing histograms and frequency polygons are also discussed in the chapter.)

10. When drawing histograms or frequency polygons to present grouped data, midpoints or exact scores are used to mark each class interval.

11. A cumulative distribution may be graphed using an **ogive** (also called an **"s" curve**). This curve is usually steep in the middle and flat at the bottom and top of the distribution.

12. The **pie graph** (or **pie chart**) looks like a circle that is divided into "wedges" or "segments." Pie graphs allow us to see relationships among the different wedges that comprise the total distribution. The size of each wedge indicates the proportion of cases in that wedge.

13. The **bar graph** (or **bar diagram**) is used to represent nominal, categorical data. It has a series of bars that do *not* touch each other that are usually ordered by height.

14. **Joint bars** may be used in bar graphs to compare data from two or more groups within each category.

15. The **line graph** is used to show trends and changes in values over time. The horizontal axis displays scores measured on a continuum (e.g., years or months). The vertical axis can be used for various types of data (e.g., test scores, sales, and income). A line connects the data points on the graphs.

16. A line graph can be used to compare multiple groups. Each group on the graph is represented by its own line.

17. The **box plot** (also called **box-and-whiskers**) consists of a box and two whiskers and is used to depict the spread and skewness of frequency distributions. The box represents the middle 50% of the distribution and the whiskers represent the top and bottom 25% of the distribution. A horizontal line inside the box represents the median of the distribution. The lengths of the whiskers show the spread of the distribution.

18. To create the box plot, the scores are first ordered and divided into four quartiles, identified as Q_1, Q_2, Q_3, and Q_4. The two middle quartiles (Q_2 and Q_3) are located *within* the box, whereas the two extreme quartiles (Q_1 and Q_4) are displayed using vertical lines (the whiskers) *outside* the box.

19. Graphs should be clearly labeled and easy to read and interpret. It is important to choose the right *scale* to draw the graphs in order to provide an accurate visual representation of the data. Other guidelines for drawing graphs are also described in this chapter.

Chapter 4

Measures of Central Tendency

A measure of central tendency is a summary score that represents a set of scores. It is a single score that is typical of a distribution of scores. There are three commonly used measures of central tendency: *mode*, *median*, and *mean*. The decision as to which of these three measures should be used in a given situation depends on which measure is the most appropriate and the most representative of the distribution.

In this chapter, we introduce the mode, median, and mean, and demonstrate how to compute them. However, you can easily obtain these statistics, along with other descriptive statistics, by using readily available computer programs.

Mode

The mode of a distribution is the occurrence with the greatest frequency in that distribution. For example, let's look at the number of hours 8 employees spent on the Internet in Table 4.1 and examine the frequency column. We can see that the score of 8 is repeated the most (3 times); therefore, the mode of the distribution is 8.

Table 4.1   Number of Hours Spent on the Internet by 8 Employees

Internet Hours	Frequency
9	1
8	3
6	2
5	2

A score should repeat at least twice in order to be considered a mode. If two scores have the same frequency, the distribution is called bimodal and if no score repeats more than one time, the distribution is classified as amodal (has no mode). If three or more scores repeat the same number of times, the distribution is referred to as a multimodal distribution. Table 4.2 shows an example of a *bimodal* distribution.

Table 4.2	A Bimodal Distribution with Two Modes (4 & 5)

Score	Frequency
6	1
5	2
4	2
3	1
2	1

In a frequency polygon, the mode is the peak of the graph (Figure 4.1, *Part a*). In a bimodal distribution, there are two peaks, both the same height (Figure 4.1, *Part b*).

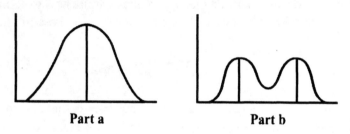

Part a   Part b

Figure 4.1   Frequency polygons with one mode (Part a) and two modes (Part b)

Median

The median is a middle point of a distribution of scores that are ordered. Fifty percent of the scores are above the median, and 50% are below it. For example, Table 4.3 shows the distribution of employees' incomes in a department. According to this table, the income of $60,000 is the median because there are 3 employees' salaries listed above it and 3 below it.

Table 4.3	Income of 7 Employees

Income
$100,000
80,000
70,000
60,000
40,000
20,000
10,000

The median is a *point* and it does not have to be an actual score in that distribution. For example, suppose a company hires employees in each of 4 different months. For each of the 4 months the company hired 10, 8, 7, and 6 employees. The median of the distribution is 7.5 employees, even though the company only hired whole numbers of employees.

When the *exact* median is hard to calculate, it can be *estimated*. For example, the median of the 7 scores of a college football team in Table 4.4 is estimated to be 8, even though the number of scores above is not exactly the same as the number of scores below it. As we can see, there are 3 scores above the median of 8 and 2 scores below it.[1]

Table 4.4	Football Scores With a Median of 8

Score
13
11
9
8
8
7
6

Mean

The mean, which is also called the arithmetic mean, is obtained by adding up the scores and dividing that sum by the number of scores. The mean is sometimes called "the average," although the word *average* may also be used in everyday life to mean "typical" or "normal." The mean, which is used in both descriptive and inferential statistics, is used more often than the mode or the median.

The statistical symbol for the *mean of a sample* is (pronounced "ex bar") and the symbol for the *population mean* is μ, the Greek letter *mu* (pronounced "moo" or "mew"). The statistical symbol for "*sum of*" is Σ (the capital Greek letter *sigma*). A *raw* score is represented in statistics by the letter X. A **raw score** is a score as it was obtained on a test or any other measure, without converting it to any other scale. "ΣX" means "the sum of all the X scores."

The mean serves as the best measure when we have to estimate an unknown value of any score in a distribution (for both samples and populations). That is, if the exact value of a particular score is unknown, the mean may be used to estimate that score.

While the mean is often the best measure of central tendency, there are instances where the mean is not an accurate representative score. To illustrate this point, let's look at Figure 4.2. *Part a* shows a *symmetrical* bell-shaped distribution, where the majority of the scores cluster around the mean. In that distribution, the mean serves as an appropriate representative score.

---

1. **A HINT:** Computer programs can provide the *exact* median of each distribution. There are also ways to compute the exact median by hand.

Part b in Figure 4.2 depicts a *negatively skewed* distribution and *Part c* depicts a *positively skewed* distribution. As we can see, in these skewed distributions, the mean is pulled toward the "tail" and it does not represent a point around which scores tend to cluster. Note that in a positively skewed distribution the scores tend to cluster *below* the mean, and in a negatively skewed distribution the scores tend to cluster *above* the mean.

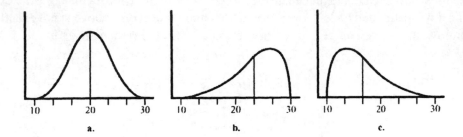

Figure 4.2   Symmetrical distribution (*Part a*), negatively skewed distribution (*Part b*), and positively skewed distribution (*Part c*)

In research, the mean of a sample is often used to estimate the population mean (μ). In many studies, researchers are interested in finding the mean of the population; however, it may not be practical or possible to study the whole population in order to find its mean. Therefore, they select a sample, measure it to obtain its mean, and use that mean to estimate the population mean.

Comparing the Mode, Median, and Mean

The mean is affected by *every* score in the distribution, because in order to calculate the mean, all the scores are first added before dividing that sum of scores (ΣX) by the number of scores. Changing even one score in the distribution results in a change in the mean. By contrast, the mode and the median may or may not be changed as a result of a change in one score. This characteristic of the mean can be both an advantage and a disadvantage. It is an advantage because the mean is a measure that reflects every score in the distribution. However, it is a disadvantage when there are extreme scores in a skewed distribution. Let's look, for example, at these 6 scores: 10, 12, 13, 13, 15, and 16. The mode is 13, the median is 13, and the mean is 13.17. All three measures are similar to each other and all can represent the distribution. Now, let's change the last score from 16 to another score, such as 40. This change has no impact on the mode or the median, but the mean changes drastically from 13.17 to 17.17. The extreme score of 40 in the second distribution "pulled" the mean upward. Consequently, the mean of 17.17 does not represent any of the 6 scores in the second distribution. It is too high for the first 5 (10, 12, 13, 13, and 15) and is much too low for the last score of 40.

Not all three measures of central tendency can be used with all types of data. Mode is the only measure of central tendency that can be used with nominal scale data. As you recall, in nominal scales the observations are not ordered in any way (see Chapter 2). Because the mode is an index of frequency, it can be used with observations that are not ordered. Mode can also be used with data measured on ordinal, interval, and ratio scales.

To find the median (the middle point), we need to be able to order the scores. A nominal scale has no order; therefore, the median cannot be used with nominal scale data. However, median can be used with ordinal, interval, and ratio scales where scores can be ordered. The mean can be computed only for interval and ratio scale data because to calculate the mean, we need to add the scores and divide the sum by the number of scores.

The mode and the median are most often used for descriptive statistics, whereas the mean is used for descriptive statistics *and* inferential statistics. For example, the mean is used to compute the variance, standard deviation, and *z* scores (see Chapters 5 and 6). It can also be used to compute other statistical tests such as the *t* test (see Chapter 10).

When distributions of scores that are measured on interval or ratio scales include extreme scores, the median is usually chosen as a measure of central tendency. For example, assume we have a few expensive homes in an area where most of the homes are moderately priced. The mean housing price may cause potential buyers, who rely on this information, to think that the houses in that area are too expensive for them. Therefore, when the mean is inflated, it does not serve as a true representative score, and the median should be used.

Summary

1. A **measure of central tendency** is a summary score that represents a set of scores. There are three commonly used measures of central tendency: *mode*, *median*, and *mean*.

2. The **mode** of a distribution is the score that occurs most frequently in that distribution.

3. A distribution of scores may have one mode, two modes (**bimodal**), three or more modes (**multimodal**), or no mode (**amodal**).

4. The **median** is the middle point of a distribution of scores that are ordered. Fifty percent of the scores are above the median and 50% are below it.

5. The **mean**, which is also called the **arithmetic mean**, is calculated by dividing ΣX (the total sum of the scores) by the number of scores (N or n). The symbol for the population mean is μ (the Greek letter mu) and the symbol for the sample mean is ("ex bar").

6. The *mean* serves as the best measure when we have to estimate an unknown value of any score in a distribution (for both samples and populations).

7. In research, the mean of a *sample* () is often used to estimate the *population* mean (μ).

8. The mean is affected by *every* score in the distribution, because to calculate the mean, all the scores are first added before dividing that sum of scores (ΣX) by the number of scores. Changing even one score in the distribution results in a change in the mean. By contrast, the mode and the median may or may not be changed as a result of a change in one score.

9. The mode can be used with nominal, ordinal, interval, and ratio scales. The median can be used with ordinal, interval, and ratio scales. The mean can be used with interval and ratio scales.

10. The mean is not an appropriate measure of central tendency when interval or ratio scale distributions have extreme scores, because it may yield a skewed measure. In such cases, the median, which is not affected by extreme scores, should be used.

11. The mode and the median are most often used for descriptive statistics, whereas the mean is used for descriptive statistics *and* inferential statistics. The mean can also be used to compute other statistical tests such as the *t* test.

Chapter 5

Measures of Variability

We have described a measure of central tendency (a mode, a median, or a mean) as a representative score; that is, a single number that represents a set of scores. These measures indicate the center of the distribution. The graphs in Figure 5.1 illustrate that, in addition to central tendency measures, it is often necessary to obtain an index of the *variability* or *spread* of the group. Let's look at Figure 5.1 and study *Part a*, which shows two groups, Group A and Group B.

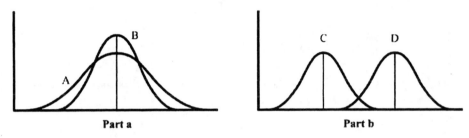

Part a   Part b

Figure 5.1   A graph showing two groups with the same means, but different spreads (*Part a*); and a graph showing two groups with different means, but the same spreads (*Part b*)

Note that the two groups share the *same mean*, but Group A has a larger spread (i.e., Group A is more heterogeneous than Group B). Group B seems to be more homogeneous and this group's scores tend to cluster closer to the mean.

Next, examine *Part b* of Figure 5.1. Notice that Group C and Group D have the same spread, or variability, but Group D has a higher mean than Group C.[1] The examples depicted in Figure 5.1 were provided to convince you that the mean alone does not provide a complete and accurate description of a group. In addition to the mean, another index is needed in order to indicate the variability of the group. In this chapter, we discuss three measures of variability: *Range, standard deviation, and variance*.

The Range

The range indicates the distance between the highest and the lowest score in the distribution. The range is a simple and easy-to-compute measure of variability. However, it has a limited usefulness as a measure of variability and it does not give us much information about the variability *within* the distribution. The range is used much less frequently compared with the other two measures of variability discussed in this chapter (the variance and standard deviation).

---

1.   **A HINT**: How can you tell which mean is higher if there are no numbers along the horizontal axis? Remember the rule about drawing a frequency polygon (discussed in Chapter 3): The numbers increase as you move to the right on the horizontal axis. Since the mean of Group D is farther to the right, it is higher than the mean of Group C.

To illustrate why range does not tell us much about the variability within a distribution of scores, compare these two sets of numbers: 10, 10, 10, 9, 1; and 10, 2, 1, 1, 1. Both have the same range, yet they represent a very different combination of scores.

Standard Deviation and Variance

The distance between each score in a distribution and the mean of that distribution (X-X) is called the deviation score.[2] Looking at Figure 5.2, we can expect higher deviation scores in *Part a*, where the scores are widely spread, than in *Part b*, where most of the scores cluster around the mean. Therefore, it would make sense to be able to calculate an index of the mean (average) of the deviation scores. We would expect that in distribution with a high spread of scores, the index value would be higher than in distributions where most scores are closer to the mean.

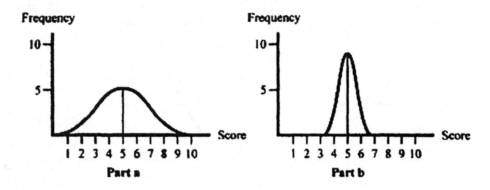

Figure 5.2   Two graphs with the same means but different spreads

The mean of the deviation scores is called the **standard deviation**, abbreviated often as SD. The SD describes the mean distance of the scores around the distribution mean. Squaring the SD gives us another index of variability, called the **variance**. As you will next see, the variance is needed in order to calculate the SD.

Let's look at the computational steps needed to calculate the deviation scores, variance, and SD. In order to simplify the computations, we will use as an example a small distribution of 5 numbers. The numbers are 6, 5, 4, 3, and 2 (see Table 5.1). Our goal is to find the mean of the distances between the scores and their mean (i.e., the mean of the deviation scores).

To compute the deviation scores, we need to first compute the mean of the 5 scores. We do so by adding up the raw scores (ΣX) and dividing them by the number of scores ($n=5$). The mean is 4.

$$\text{Mean} = \overline{X} = \frac{\Sigma X}{n} = \frac{20}{5} = 4$$

---

2.   **A HINT**: Some textbooks use the symbol x (lower case "ex") to represent the deviation score. Other texts, including this textbook, X-X to represent the deviation score.

Table 5.1   **Computing Deviation Scores**

(Raw Scores) X	(Deviation Scores) $X - \bar{X}$
6	6 − 4 = 2
5	5 − 4 = 1
4	4 − 4 = 0
3	3 − 4 = -1
2	2 − 4 = -2
$\Sigma X = 20$	$\Sigma(X - \bar{X}) = 0$
Sum of Raw Scores	**Sum of Deviation Scores**

As you can see in Table 5.1, the sum of the deviation scores is 0 (zero). Therefore, we cannot divide that sum by 5 (the number of scores) to find the mean of the deviation scores, as we were planning to do. As a matter of fact, with any combination of numbers, the mean of the deviation scores will always be 0 (zero). However, we can overcome this problem by squaring each deviation score first, then finding the mean of the *squared* deviation scores.

The next step, then, is to square each deviation score, add up the squared deviation scores, and find their mean (see Table 5.2). The mean of the squared deviations is the **variance**. Therefore, the variance may be defined as the *mean of the squared deviations around the mean*. The statistical symbol used to represent the sample variance is S^2.

It is important to know whether our distribution of scores is a *sample* or a *population* because there is a difference in the computational steps for finding the variances of samples and populations. The difference is in the denominator (see the computations that follow Table 5.2). The denominator is $n-1$ (number of scores minus 1) for samples, and N (number of scores) for populations. Because there are 5 scores in Tables 5.1 and 5.2, we assume the scores to be a *sample* rather than a *population*.

Table 5.2   **Computing the Variance**

X	$X - \bar{X}$		$(X - \bar{X})^2$
6	6 − 4 =	2	4
5	5 − 4 =	1	1
4	4 − 4 =	0	0
3	3 − 4 =	-1	1
2	2 − 4 =	- 2	2
$\Sigma X = 20$	$\Sigma(X - \bar{X}) = 0$		$\Sigma(X - \bar{X})^2 = 10$

The formula for the computation of the sample variance (S^2) is:

$$\text{Variance} = S^2 = \frac{\Sigma(X - \overline{X})^2}{n-1}$$

Where	S^2	=	Sample variance
	$\Sigma(X - \overline{X})^2$	=	Sum of the squared deviations from the mean
	n-1	=	Number of scores minus 1

Replacing the symbols with our numbers from Table 5.2 will give us a sample variance of 2.5.

$$\text{Variance} = S^2 = \frac{\Sigma(X - \overline{X})^2}{n-1} = \frac{10}{4} = 2.5$$

Our reason for computing a measure of variability was that we would find an index of the mean (average) of the deviations around the mean. Clearly, the variance is not a good measure for that purpose. In our example, we computed a variance of 2.5, which is not representative of the deviation scores of 2, 1, 0, -1, and -2 (see Table 5.1). The reason why the variance is higher than all the deviation scores in our example is pretty obvious: As you recall, the sum of the deviation scores in Table 5.1 was 0. To overcome this problem, we *squared* the deviation scores (see Table 5.2). Therefore, the variance is higher than all the deviation scores. To get over the problem of obtaining such an inflated index, we simply reverse the process. In other words, we find the *square root* of the variance, which would help us get back to the original units we used in Table 5.1. The square root of the variance is the *standard deviation* (SD). The computation formula for the standard deviation is:

$$S = \sqrt{\text{VAR}} = \sqrt{S^2}$$

Where	S	=	Sample standard deviation
	S^2	=	Sample variance

In our example, the variance is 2.5 and its square root, the SD, is 1.58.

$$S = \sqrt{\text{VAR}} = \sqrt{S^2} = \sqrt{2.5} = 1.58$$

Conversely, we can reverse the process and find the variance by *squaring* the SD. For example, when the SD is 5, we can find the variance (S^2) by squaring the SD.

$$S^2 = SD^2 = 5^2 = 25$$

Where	S^2	=	Sample variance
	SD^2	=	Sample SD, squared

Because the SD is the square root of the variance, it is usually *smaller* than the variance. However, this is not always the case. As you know, the square root of 1 is 1. In fact, when the variance is less than 1, the SD will be *higher* than the variance. For example, when the variance is 0.8, the SD is 0.89 and when the variance is 0.5, the SD is 0.7.

As we said, the symbol that is used for the *sample* standard deviation is S and the one used for the sample variance is S^2. By comparison, Greek letters are used for the population values. The symbol used for the *population* standard deviation is σ (Greek lower case letter *sigma*) and the symbol for the population variance is σ^2. There is a difference in the equations used to compute the sample and population variances and standard deviations. Following is an explanation of that difference.

Computing the Variance and *SD* for Populations and Samples

The standard deviation of a population (σ) is a fixed number, but the sample standard deviation (S) varies, depending on the sample that was selected. If we select several samples from the same population and compare the standard deviations of such samples, we are likely to see that not all of them are exactly the same.[3] When researchers started comparing such samples to the population from which they were selected, they realized that not all samples have the same means and standard deviations. Further, researchers also found that variances and standard deviations from samples were consistently *lower* than the variances and standard deviations of the populations from which the samples were selected. This was especially true with small samples (with $n \leq 30$). In conducting research, researchers often use the standard deviation and variance from a single sample to *estimate* the variance and standard deviation of the population. Because the sample variance and standard deviation are likely to consistently *under*estimate the population variance and standard deviation, there was a need to modify the equations used to compute the variance and standard deviation. These modified equations result in a slightly higher variance and standard deviation that are more representative of the variance and standard deviation of the population.

For example, in Table 5.2 we have only 5 scores. It is very likely that such a small group of scores is a sample, rather than a population. Therefore, we computed the variance and SD for these scores treating them as a *sample* and used a denominator of n-1 in the computations that followed Table 5.2. When, on the other hand, we consider a set of scores to be a population, we should use a denominator of N to compute the variance.

The difference in the choice of denominator is especially important when the number of cases in the distribution is small. Consider, for example, the data in Table 5.2. In our computation of the variance, which follows that table, we had a numerator of 10 and used a denominator of 4 (n-1). If we had decided that the group of 5 scores is a population, we should have used a denominator of N when computing the variance. Consequently, with a numerator of 10, we would have obtained a variance of 2 instead of our variance of 2.5, and a SD of 1.41 instead of our SD of 1.58.

Clearly, when the sample is large, the choice of the proper denominator (N vs. n-1) to be used for the computation of the variance and SD is not as important as when the sample is very small. For example, let's assume we have a set of 100 scores and the numerator in the formula used to compute the variance is 800. If that set of scores is considered a *population*, then to compute the variance we will divide the numerator by 100. The variance would be 8.00 (800 divided by 100). The SD for that population of 100 scores would be 2.83 (the square root of the variance). If, on the other hand, the set of scores is considered to be a *sample*, then the denominator would be 99 (n-1) and the variance would be 8.08 (800 divided by 99). The SD would be 2.84 (the square root of 8.08). Clearly, there is very little difference between the two standard deviations (2.83 for a population and 2.84 for a sample) when the sample size is large.

---

3. **A HINT:** The same applies to means: means from different samples selected from the same population are likely to differ, whereas the mean of the population is a fixed number (see Ch. 2).

Using the Variance and *SD*

In real life, you are not likely to have to calculate either the variance or the SD by hand. There are many computer programs that are easy to use that can calculate both the variance and SD. We introduced the computation steps here simply to explain these concepts.

A higher variance may show higher variability in a group, compared with a lower variance. However, these values are difficult to interpret, because they are measured in *squared* units. The SD, on the other hand, is measured in the same units as the original data, and is easier to interpret. For example, when measuring age in years, a SD of 3 means that, on the average, the ages of the members of the group deviate 3 years from the mean.

The variance may be used as an intermediate step in the computation of the standard deviation. The variance is also found in the computational steps of some statistical tests such as *t* test (see Chapter 10), and analysis of variance (ANOVA) (see Chapter 11). Standard deviation, on the other hand, is often used to summarize data, along with the mean or other measures of central tendency. For example, in reporting results of tests, we are most likely to use summary scores, such as the mean and standard deviation. Technical manuals for tests show an extensive use of the mean and SD. Further, scales of tests are usually described in terms of their mean and SD. For example, we are told that a certain IQ test has a mean of 100 and a SD of 15. (These concepts are presented in Chapter 6, which discusses the normal curve, and Chapter 7, which discusses test scores.)

Variance and SD in Distributions with Extreme Scores

The variance and the SD are sensitive to extreme scores. Having skewed distributions with even one extreme score may substantially increase the variance and SD. Consider these two sets of scores, Set 1 and Set 2 (Table 5.3). Note that the two sets are the same with the exception of one extreme score in Set 2 (40, the first score in Set 2).

Table 5.3	Two Sets of Scores, With An Extreme Score in Set 2	
	Set 1	Set 2
	11	40
	10	10
	9	9
	8	8
	7	7
	6	6
	5	5
	4	4
	3	3
	2	2
ΣX =	65.00	94.00
$\overline{X}$ =	6.50	9.40
VAR =	9.17	122.27
SD =	3.03	11.06

Set 1 has a SD of 3.03, which seems like a good representation of the distances of the scores around their mean of 6.50. The mean in Set 1 is also a good representation of the scores in that set. By comparison, the variance and SD in Set 2 are much higher than those in Set 1, due to the extreme score of 40. The SD is supposed to be an index of the average distances of the scores around their mean. In Set 2, the SD of 11.06 provides misleading information. It implies a much higher variability in the group, compared with the SD in Set 1. In fact, all the scores in Set 2, with the exception of the first score of 40, are fairly close together. The mean of 9.40 in Set 2 is also misleading and it is higher than most of the scores in that set.

Factors Affecting the Variance and SD

As you probably have noticed, there is a relationship between the range, variance, and SD; the wider the range, the higher the variance and SD. The range is higher when the group is more heterogeneous regarding the characteristic being measured. This characteristic can be, for example, income, sales, expenses, or age. The range, variance, and SD are also higher when there is at least one extreme score in the distribution, even if the rest of the scores cluster together (see Set 2 in Table 5.3). These three measures (range, SD, and variance) tend to be lower when the scores cluster together, as is the case in Set 1 in Table 5.3.

In training and academic applications, the length of a test can affect the variance and SD. A longer test has the potential to spread the scores more widely and have a higher SD than does a shorter test. Compare, for example, two tests: Test A, with 100 items, and Test B, with 10 items. Let's assume the mean of Test A is 50, and the mean of Test B is 5. In test A, people might score up to 50 points above or below the mean. In Test B, on the other hand, people might score only up to 5 points above or below the mean. Since the SD is a measure of the mean (average) distance of the scores from the mean, the SD is likely to be higher in Test A than in Test B. The same would apply to the variance of the distribution, which is obtained by squaring the SD of that distribution. Variances of short tests tend to be smaller than variances of long tests.

Another factor that can affect the variance and SD in testing situations is the level of difficulty of the test. When a test is very easy, most respondents answer all the questions correctly; therefore, the scores cluster together and there is little variability in the scores. Consequently, the variance and SD are likely to be lower. Similarly, scores from tests that are very difficult for all examinees tend to cluster together at the low end of the distribution. When scores cluster at the high end or the low end of the distribution curve, the variance and SD tend to be lower than in cases where the scores are spread along a bell-shape distribution. Norm-referenced tests are designed to spread the scores widely to create a bell-shape distribution (see Chapter 7). Scores from such tests would have higher variance and SD than those obtained on tests where the scores tend to cluster together.

Summary

1. To describe a distribution of scores, an index of **variability**, as well as a measure of central tendency, is needed.

2. The **range** is the distance between the highest and the lowest scores in the distribution. To calculate the range, subtract the lowest score from the highest score.

3. The range is an index of the variability of the group, and it is used mostly for descriptive purposes.

4. The **deviation score** is the distance of the raw score from the mean, indicated by $X -$ (i.e., the score minus the mean).

5. The sum of the deviation scores (the distances between the raw scores and the mean of that distribution) is always 0 (zero).

6. The **variance** is the mean of the squared deviations. To calculate it, square each deviation score, add all the squared deviations, and divide their sum by n-1 (the number of scores minus 1) for the *sample* variance.

7. In the equation used to compute the *population* variance, the denominator is N (number of scores). Choosing the correct equation is especially important when the sample size is small ($n \leq 30$).

8. The **standard deviation** (SD) is the mean (average) distance of scores from the mean. It can be computed by finding the square root of the variance $SD = \sqrt{Variance}$

9. When we square the SD, we can find the variance (Variance = SD^2).

10. The symbol that is used for the *sample* variance is S^2 and the symbol used for the *population* variance is σ^2. The symbol for the *sample* standard deviation is S and the symbol for the *population* standard deviation is σ (the Greek lower case letter *sigma*).

11. The standard deviation of a population (σ) is a fixed number, but the sample standard deviation (S) varies, depending on the sample that was selected.

12. Empirical research has shown that the sample variance and standard deviation consistently *under*estimate the population variance and standard deviation. Therefore, the equations used to compute the variance and standard deviation of samples were modified to produce slightly higher variance and standard deviation that are more representative of the variance and standard deviation of the population.

13. The standard deviation is measured using the same units as the original data, and is easier to interpret than the variance. Standard deviation is often used along with the mean in summarizing and reporting test data.

14. The variance is not commonly used when describing a distribution of scores. The reason is that the variance, which is expressed in *squared* units, tends to be much larger than the majority of the deviation scores around the mean of the distribution.

15. The range, variance, and standard deviation are sensitive to extreme scores. Using the same test, groups with a wide range of scores (heterogeneous groups), have higher ranges, variances, and standard deviations than groups where scores cluster together (homogeneous groups).

16. Variance and standard deviation of longer tests (with more items) tend to be higher than the variance and standard deviation of shorter tests (with fewer items).

17. In tests that are very easy or very difficult, the scores of examinees tend to cluster in one end and the variance and standard deviation are likely to be low on such tests.

18. In norm-referenced tests, the examinees' scores are usually spread along a bell-shaped curve. Therefore, the scores tend to have a wider range resulting in higher variance and standard deviation (see Chapter 7).

Part Three

The Normal Curve and Standard Scores

Chapter 6

The Normal Curve and Standard Scores

The Normal Curve

For years, scientists have noted that many variables in the behavioral and physical sciences are distributed in a bell shape. These variables are *normally distributed* in the population and their graphic representation is referred to as the **normal curve**.[1] For example, in the general population, the mean of the most commonly used measure of IQ is 100. If the IQ scores of a group of 10,000 randomly selected adults are graphed using a frequency polygon, the graph is going to be bell-shaped, with the majority of people clustering just above or below the mean of 100. There would be increasingly fewer and fewer IQ scores toward the right and left tails of this distribution. Similarly, if we were to record the salaries of 1,000 senior level executives from different industries and then graph the data, we would see that it forms a normal curve.

The development of the mathematical equation for the normal distribution is credited, according to some sources, to the French mathematician Abraham Demoivre (1667-1754). According to other sources, it was the German mathematician Karl Friedrich Gauss (1777-1855) who developed the equation. Thus, the normal curve is also called the "Gaussian Model."

The normal curve is a theoretical, mathematical model that can be represented by a mathematical formula. However, since many behavioral measures are distributed in a shape like the normal curve, the model has practical implications in the behavioral sciences. In this chapter, we show how this model can be applied to business.

The **normal distribution** is actually a group of distributions, each determined by a mean and a standard deviation. Some of these distributions are wider and more "flat", while others are narrower, with more of a "peak" (see Figure 6.1).[2]

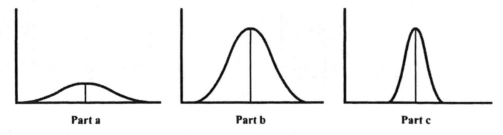

Part a   Part b   Part c

Figure 6.1   Three normal distributions with different levels of "peakedness"
(or "flatness")

---

1. **A HINT:** The standard normal curve is referred to by most people as the normal curve, or a bell-shaped distribution.
2. **A HINT:** The level of "peakedness" or "flatness" of the curves is called *kurtosis*.

Regardless of the exact shapes of the normal distributions, all share four characteristics:

1. The curve is symmetrical around the vertical axis (half the scores are on the right side of the axis, and half the scores are on its left).
2. The scores tend to cluster around the center (i.e., around the mean, or the vertical axis).
3. The mode, median, and mean have the same values.
4. The curve has no boundaries on either side (the tails of the distribution are getting very close to the horizontal axis, but never quite touch it).[3]

Although many characteristics are normally distributed, measuring and graphing these characteristics for a small number of cases will not necessarily look like the normal curve. *Part a* in Figure 6.2 depicts a distribution of scores obtained for a smaller sample and *Part b* depicts scores from a larger sample. Note that the graph of the distribution in *Part b* looks "smoother" than the graph of the distribution in *Part a*. As the number of cases increases, the shape of the distribution is more likely to approximate the normal curve.

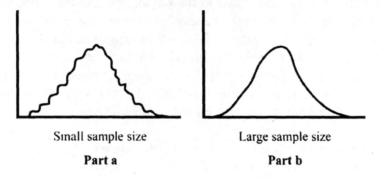

Small sample size　　　　Large sample size

Part a　　　　　　　**Part b**

Figure 6.2　Normal curve distributions with a smaller sample size (Part a) and a larger sample size (Part b)

The normal curve is divided into segments and each segment contains a certain percentage of the area under the curve (see Figure 6.3). The distances between the various points on the horizontal axis are equal, but the segments closer to the center contain more scores than the segments farther away from the center.

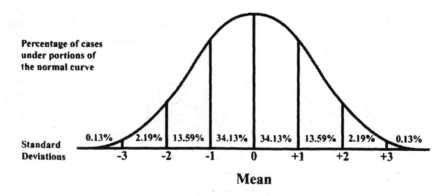

Figure 6.3　A bell-shaped (normal) distribution with a mean of 0 and a standard deviation of 1

3. **A HINT:** Keep in mind that this is a theoretical model. In reality, the number of scores in a given distribution is finite, and certain scores are the highest and the lowest points of the distribution.

Figure 6.3 shows the normal curve with a distribution of scores that have a mean of 0 and a standard deviation (SD) of 1. The units *below* the mean (on the left side) are considered *negative* (e.g., -1, -2), and the units *above* the mean (on the right side) are considered *positive* (e.g., +1, +2).

In normal distributions, 34.13% of the scores are expected to be between the mean and +1*SD* and the area between the mean and 2*SD* above the mean is expected to include 47.72% (34.13 + 13.59 = 47.72) of the scores (see Figure 6.4). The area between the mean and −1SD contains 34.13% of the scores and the area between the mean and −2SD includes 47.72% of the scores.The area between 3*SD* above and 3*SD* below the mean is expected to contain almost all the cases in the distribution, 99.74%.

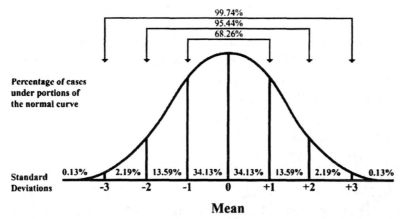

Figure 6.4   The percentages of the area under the normal curve at ±1*SD*, ±2*SD*, and ±3*SD*

The normal curve can be used to describe, predict, and estimate many types of variables that are normally distributed. If we know the distribution mean and standard deviation, we can estimate the percentages of scores in different parts of the distribution.

For example, we can use the normal curve to estimate the number of minutes a person stands in line before being helped at an airline ticket counter by using the information about the areas under the normal curve. Suppose we know the mean time for standing in line is 10 minutes with a standard deviation of 2. Using the information in Figure 6.4, we can mark 12 minutes as the point on the normal distribution graphs at +1*SD* (a mean of 10 plus 2 points for 1 standard deviation) and the point at +2*SD* as 14 minutes. In the area below the mean, -1*SD* corresponds to 8 minutes in line and the point at −2*SD* corresponds to 6 minutes. We can determine that in the general population, approximately 34% of the people are expected to have to wait between 10 and 12 minutes (between the mean and +1*SD*) before reaching the ticket counter. We can also use Figure 6.4 to learn that we can expect about 68% (or two thirds) of the people in the population to have to stand in line between 8 and 12 minutes. In various distributions, the area between ±1*SD* is considered within the normal range.) The airline may know that if customers have to wait more than +2*SD* above the mean number of minutes (i.e., more than 14 minutes) that these people will complain about having to wait too long in line and may consider using another airline the next time they fly.

Standard Scores

Until now, two types of scores were introduced in the book: *individual scores* (raw scores) and *group scores* (mode, median, mean, range, variance, and standard deviation). Raw scores are scores obtained by individuals on a certain measure and group scores are summary scores that are obtained for a group of scores. However, both types of scores are scale specific and cannot be used to compare scores on two different measures, each with its own mean and standard deviation. To illustrate this point, let's look at the following example.

Suppose we want to compare the scores obtained by a prospective MBA student on two parts of the Graduate Management Admission Test (GMAT): Verbal and Quantitative. Let's say that the student received a score of 40 on the Verbal part of the test and 27 on the Quantitative section. Because the two tests are different we cannot conclude that the student performed better in the Verbal section than the Quantitative portion of the test. Knowing the student's score on each test will not allow you to determine on which test the student performed better. We do not know, for example, how many items were on each test, how difficult the tests were, and how well the other students did on the tests. Simply put, the two tests are not comparable.

To be able to compare scores from different tests, we can first convert them into standard scores. A **standard score** is a derived scale score that expresses the distance of the original score from the mean in standard deviation units. Once the scores are measured using the same units, they can then be compared to each other. Two types of standard scores are discussed in this chapter: *z scores* and *T scores*.[4]

z Scores

The *z* score is a type of standard score that indicates how many standard deviation units a given score is *above* or *below* the mean for that group. The *z* scores create a scale with a mean of 0 and a standard deviation of 1. The shape of the *z* score distribution is the same as that of the raw scores used to calculate the *z* scores.

The theoretical range of the *z* scores is $\pm\infty$ (plus/minus infinity). However, since the area above a *z* score of +3 or below a *z* score of -3 includes only 0.13% of the cases, for practical purposes most people use only the scale of -3 to +3. To convert a raw score to a *z* score, the raw score as well as the group mean and standard deviation are used. The conversion formula is:

$$z = \frac{X - \overline{X}}{S}$$

Where X =   Raw score
          $\overline{X}$ =   Group mean
          S =   Group standard deviation (SD)

Table 6.1 presents the raw scores (out of 60 possible points) of one test taker on the four sections (Verbal, Quantitative, Writing Assessment, Total) of the new GMAT test. The table also displays the means and standard deviations of all who were the first to take the new test. The computations that follow the table demonstrate the process for converting raw scores into *z* scores.

---

4.   **A HINT:** The *T* scores are not related to the *t* test that is discussed in Chapter 10.

The raw scores that are *above* the mean convert into *positive z* scores and the raw scores *below* the mean convert into *negative z* scores. Consequently, about half the examinees are expected to get positive *z* scores and half are expected to get negative *z* scores. As we can see in Table 6.1, a test taker may answer many questions correctly (e.g., see the score of 55/60 on the Verbal test) yet get a *z* score of 1.67. It is clear that for reporting purposes, *z* scores are not very appealing. Additionally, many students would obtain negative *z* scores (see, for example, the student's score of -0.50 on the Quantitative section and -1.60 on the Writing section) even though the student answered correctly quite a few questions (27 and 25, respectively).

Table 6.1   GMAT Raw Scores of One Student: Student's Raw Scores, GMAT Mean Scores and Standard Deviation, and Student's *z* Scores

Section	Raw Score	Mean	SD	z Score
Verbal	55	50	3	$\dfrac{55-50}{3}=+1.67$
Quantitative	27	31	8	$\dfrac{27-31}{8}=-0.50$
Writing	25	41	10	$\dfrac{25-41}{10}=-1.60$
Total	42	30	6	$\dfrac{42-30}{6}=+2.00$

T Scores

The *T* score is another example of a standard score measured on a scale with a mean of 50 and a *SD* of 10 (Figure 6.5). In order to calculate *T* scores, *z* scores have to be calculated first. Using this standard score overcomes the problems associated with *z* scores. All the scores on the *T* score scale are positive and range from 10 to 90. Additionally, they can be reported in whole numbers instead of decimal points.

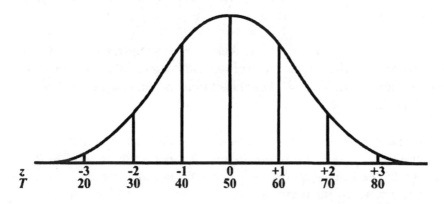

Figure 6.5   A normal curve showing *z* scores and the corresponding *T* scores

In order to convert z scores to T scores, we multiply each z score by 10, and add a constant of 50 to that product. The formula to convert from a z score to a T score is:

$$T = 10(z) + 50$$

To illustrate the conversion process, let's convert the z scores in Table 6.1 into T scores using the conversion formula. The computations are displayed in Table 6.2. T scores are usually rounded off and reported as whole numbers. Inspecting the table, we can see that negative z scores convert to T scores that are below 50. For example, a z score of -0.50 is converted to a T score of 45. Positive z scores convert to T scores that are higher than 50. For example, a z score of +2.00 is converted to a T score of 70.

Table 6.2   **Conversion of z Scores to T Scores**

Test	z Score	T Score
Verbal	+1.67	10 (+1.67) + 50 = 67.0 or 67
Quantitative	-0.50	10 (-0.50) + 50 = 45.0 or 45
Writing	-1.69	10 (-1.69) + 50 = 33.1 or 33
Total	+2.00	10 (+2.00) + 50 = 70.0 or 70

Normal Curve and Percentile Ranks

A **percentile rank** of a score is defined by most people as the percentage of examinees that scored *at* or *below* that score. For example, a percentile rank of 65 (P_{65}) means that 65% of the examinees scored at or below that score. Other definitions of a percentile rank state that it indicates the percentage of examinees that scored *below* that score (omitting the word "at"). The second definition is the one used most often by commercial testing companies on their score reports.

In practice, a percentile rank of 100 is not reported. We cannot say that a person with a certain raw score did better than 100% of the people in the group, because that person has to be included in the group. Instead, 99% (or in some cases, 99.9%) is considered the highest percentile rank.

Percentiles are used to describe various points in a distribution. For example, a percentile rank of 70 (P_{70}) is said to be at the 70th percentile. Since percentiles represent an ordinal, rather than interval or ratio scale, they should not be manipulated (e.g., added or multiplied). If manipulation is desired, percentiles should first be converted to z scores (which have equal intervals) or to raw scores.

The normal curve can be used to calculate percentiles assuming that the distribution of scores is normally distributed (see Figure 6.6). For example, a z score of +1 corresponds to a percentile rank of 84.14 (or 84). We find that percentile rank by adding up the percent of scores between the mean and a z score of +1 on the normal curve (it is 34.14%) to the percent of scores below the mean (50%). A z score of -2 corresponds to a percentile rank of 2 (the percent of area under the normal curve below a z score of −2).

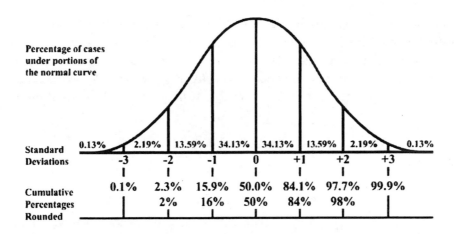

Figure 6.6   The normal curve with standard deviations and cumulative percentages

In real life, obtained z scores are not likely to be whole numbers, but rather numbers such as +1.67 or -1.69 (see Table 6.2). A special table that lists z scores and the area under the normal curve corresponding to these scores may be used in such cases. The percentile score which corresponds to each z score can be found by calculating the area under the normal curve corresponding to that z score.

To use the table like the one from which Table 6.3 below is taken, you have to locate the z score in the left hand column (column 1). Moving to column 2 marked "*Area from Mean to z,*" locate the number corresponding to the z score. This number indicates the area between the mean and the z score. The 4 decimals can be converted to percentages by rounding down to 2 digits. For example, .4574 is converted to 46%. For positive z scores, we add these percentages to 50%, and for negative z scores, we subtract these percentages from 50%.

In Table 6.3 we take a z score of +1.67; the corresponding number in column 2 is .4525, which is converted to 45% and added to 50%, resulting in a percentile of 95. For a z score of -0.50, the corresponding number of .1915 is converted to 19%, which we then subtract from 50% (because the z score is negative), resulting in a percentile of 31. For a z score of -1.69, we subtract 45% from 50%, to get a percentile of 5; and a z score of +2.00 is converted to a percentile of 98 (50% + 48%).

**Table 6.3   Sample Z Scores and Percentage of Area Under the Normal Curve
Between Given Z-Scores and the Mean**

Standard *Z* Score (Column 1)	Area from Mean to Z (Column 2)
1.66	0.4515
1.67	0.4525
1.68	0.4535
1.69	0.4545

Using the normal curve to find percentiles is justified when normal distributions are studied. However, suppose a training manager wants to convert scores on a training test created by the manager into percentiles in order to report the scores to her boss. The training manager can first use the test scores to create a frequency polygon. Then, she can examine the graph to ascertain whether the distribution of scores is "normal."

In most instances where we have small group size ($n < 30$), the shape of the distribution will not approximate the normal curve. When the shape of the distribution is not assumed to be normal, it is inappropriate to use the normal curve model as a means to calculate percentiles. Instead, the training manager can use a simple formula to convert raw scores to percentile ranks. The information needed to convert a raw score to percentile rank is the number of trainees below that score, the number of trainees with the same score, and the total number of trainees. The formula is:

$$PR = \frac{N_{Below} + N_{at}}{N_{Total}}(100)$$

Where   PR   =   Percentile Rank
         N_{Below} =   Number of trainees who scored below that raw score
         N_{at} =   Number of trainees who received the same score
         N_{Total} =   Total number of trainees

For example, assume that 3 trainees in a group of 30 get a score of 25, scoring better than 21 other trainees in the class. Using this formula, we obtain a percentile rank of 80 for these trainees. In other words, they did better than, or as well as, 80% of their fellow trainees who took the same test. The computations are:

$$PR_{25} = \frac{21+3}{30}(100) = (0.80)(100) = 80$$

In practice, you are not likely to compute the percentile ranks by hand. Instead, you can use computer programs (such as SPSS) to compute percentile ranks as well as z scores.

Summary

1. The **normal curve** is the graphic representation of *normally distributed* variables in the behavioral and physical sciences.

2. The graph of the normal curve is bell-shaped, with the majority of scores clustering just above or below the mean and increasingly fewer scores at either end of the curve.

3. The normal curve is a theoretical, mathematical model that can be represented by a mathematical formula. However, since many behavioral measures are normally distributed, the model has practical implications in the behavioral sciences as well.

4. The normal distribution consists of a group of distributions, each determined by a mean and a standard devia tion. Some of these distributions are wider and more "flat" while others are narrower, with more of a "peak."

5. The **normal distribution** has four characteristics: (a) it is symmetrical around the vertical axis; (b) the scores tend to cluster around the center; (c) the mode, median, and mean have the same value; and (d) theoretically, the curve has no boundaries on either side.

6. The normal curve is divided into segments and each segment contains a certain percentage of the area under the curve. The distances between the various points on the horizontal axis are equal, but the segments closer to the center contain more scores than the segments farther away from the center.

7. In a normal distribution, 34.13% of the scores are expected to be between the mean and $1SD$. The area between the mean and $2SD$ above the mean is expected to include 47.72% of the scores, and the area between $3SD$ above and $3SD$ below the mean (written as $\pm 3SD$) is expected to contain almost all the cases in the distribution (99.74%).

8. The normal curve can be used to describe, predict, and estimate many types of variables that are normally distributed. If we know the distribution mean and standard deviation, we can estimate the percentages of scores in different parts of the distribution.

9. A **standard score** is a derived scale score that expresses the distance of the original score from the mean in standard deviation units. Standard scores, such as z scores, can be used to compare raw scores from different distributions of scores (e.g., from different tests).

10. A z score is a commonly used standard score that indicates how many standard deviation units a given score is above or below the mean for that group. The group's mean and standard deviation are used to convert the raw scores to z scores. The conversion formula is:

$$z = \frac{RawScore - Mean}{SD}$$

11. Raw scores that are *above* the mean convert into *positive z* scores and raw scores that are *below* the mean convert into *negative z* scores. Therefore, if raw scores are converted to z scores, half of the raw scores are expected to convert to positive z scores and the other half to negative z scores. Scores *exactly* at the mean, the most "average," are assigned a z score of 0.00.

12. Using z scores for the purpose of reporting students' scores can be problematic because students may be as signed negative scores, scores of 0, and scores with decimal places. Additionally, no student may get a score higher than 4.

13. T **scores** are standard scores that can range from 10 to 90, with a mean of 50 and a standard deviation of 10. To obtain T scores, it is necessary to find the z scores first. The conversion formula is: $T=10(z) + 50$.

14. A **percentile rank** of a score is defined by most people as the percentage of examinees that scored *at* or *below* that score. For example, a percentile rank of 65 (P_{65}) means that 65% of the examinees scored at or below that score. Other definitions of a percentile rank state that it indicates the percentage of examinees that scored *below* a given score (omitting the word "at"), or that it includes the percentage of scores *below* a given score *plus half* of those who have obtained that same score.

15. **Percentiles** are used to describe various points in a distribution. For example, a percentile rank of 70 (P_{70}) is said to be at the 70th percentile. Since percentiles represent an ordinal, rather than interval or ratio scale, they should not be manipulated (e.g., added or multiplied).

16. In most instances where we have small group size ($n < 30$), the shape of the distribution will not approximate the normal curve. In such cases, a simple formula to convert raw scores to percentile ranks can be used. The formula is:

$$PR = \frac{N_{Below} + N_{at}}{N_{Total}}(100)$$

Chapter 7

Interpreting Test Scores

Tests are used in all areas of life. They are often given to job applicants to determine if their personalities, skills, intelligence, and interests are a good match for the job and the organization to which they have applied. They are also used to determine admission into graduate programs, to earn professional licenses, to monitor job progress, to assess the honesty of employees, and more. Those who create tests include consultants, managers, trainers, and organizational psychologists. Some of the biggest designers of tests in business are companies that produce pre-employment tests that are designed to help organizations hire better employees, such as the *Myers-Briggs Type Indicator Instrument*. Because more than 60% of all organizations use some type of pre-employment test our discussion in this chapter will focus mostly on pre-employment tests.

There are several ways to report test scores. Some of the most common ways are *raw scores*, *percent correct*, *standard scores* (such as *z scores*), *percentile ranks*, *stanines*, *cut scores*, and *percent correct*. (See Chapter 6 for a discussion of standard scores and percentile ranks.)

It is hard to interpret raw scores obtained by an applicant on a pre-employment test if no additional information is available about the test (such as the number of items and their level of difficulty, and the scores of the other applicant-examinees who took the tests). Raw scores derived from standardized tests constructed by commercial test companies are usually converted into scale scores and norms. Managers often convert the raw scores obtained by their examinees on a manager-created test to percent correct.

Tests can be classified into two major categories: *norm-referenced* and *criterion-referenced*. The two types of tests differ in the way they are constructed and how they are used.

Norm-Referenced Tests

Norm-referenced (NR) tests include norms that allow the test user to compare the performance of an individual taking the test to that of similar examinees who have taken the test previously. These examinees comprise the *norming group*. The norming group is a sample taken from the population of all potential examinees. Stratified random sampling procedure is usually employed to select the sample used for norming. Stratification is done on characteristics such as gender, age, socioeconomic status, race, and geographic region. The norming group should be large enough and demographically represent the characteristics of the potential test takers. The test is first given to the norming group and then the scores on the test are used to generate the norms. Later, when new examinees take the test, their scores are usually compared to the scores of the norming group, rather than to the scores of others taking the test with them. However, in some cases the score of an examinee is compared to the scores of those who took the test at the same time in order to generate local norms.

In standardized tests, items have been first pilot-tested and revised, and the test administration procedures are uniform. Test items constructed for NR tests are written specifically to maximize differences among the examinees. Some items have a high level of difficulty in order to differentiate among the top examinees, while other items are easy in order to distinguish among the low-scoring examinees. Easy items may also be placed at the beginning of the test or section to encourage all test subjects. Most items are of average difficulty and are designed to be answered correctly by 30-80% of the examinees.

Commercial testing companies describe in their technical manual how the norming group was selected, its demographic characteristics, and when the norms were obtained. Other technical aspects of the test, such as its reliability and validity, are likely to be discussed in the manual as well. Norms are usually included in standardized, commercially constructed tests.

Testing companies may develop two types of norms: *national* and *local*. This is especially common with standardized employment tests, which are given to many job applicants in the U.S. In a typical organization, the human resources manager receives a computer-generated report that lists the raw scores, as well as national and local norms. The **national norms** compare the examinee to similar examinees in the population at large, while **local norms** compare the examinee to others with the same demographic characteristics in the same organization or industry.

Several tests, such as college admission tests, are designed for a particular purpose. The norming group, although more specific, is still comprised of examinees with characteristics similar to those of the potential test users. For example, the Scholastic Aptitude Test (SAT) and the ACT Assessment are normed on college-bound high school juniors or seniors. The Graduate Record Examination (GRE) is normed on students who plan to attend graduate schools. Professional graduate programs (such as business schools, law schools and medical schools) have their own admission tests that are normed on a representative sample of students who apply to these professional programs.

Test publishers report several types of scores. A typical student report includes raw scores on each subtest, as well as norms. Standardized scores, such as z and T scores, may also help place examinees performance in relation to others who took the same test. Three of the most commonly used norms are *percentiles ranks*, *stanines*, and *cut scores*.

Percentile Ranks

A *percentile rank* describes the percentage of people who scored *at* or *below* a given raw score. For example, when a raw score of 58 is converted to a percentile rank of 82, it means that an examinee with that raw score performed better than, or as well as, 82% of those in the norming group. At times, a percentile rank is described simply as the percentage of examinees that scored *below* a given score (omitting the word *at* from the definition). The standardized test publishers include percentile ranks in the reports they provide to organizations that buy and use their tests. (See Chapter 6 for a more comprehensive discussion of percentile ranks and percentiles.)

Percentile ranks are easy for managers and others to understand, which may be one of the reasons they are popular as norm-referenced scores. The public may also be familiar with the concept of percentile ranks because they are used in familiar situations such as pediatricians charting the height and weight of babies and young children.

In addition to percentile ranks, standardized test reports usually include **percentile bands.** Since the tests are not completely reliable and include a certain level of error, the band gives an estimated range of the true percentile rank. A confidence level of 68% is commonly used in constructing the band. On the test report, the band is often represented by a shaded area.

After a commercial norm-referenced pre-employment test is administered to a job applicant, the organization is likely to receive reports describing the applicant's performance on the tests. Although the format and content of the reports produced by various testing companies differ from each other, most of them provide information about the examinee's performance, including national percentile ranks and percentile bands on the traits covered by the test.

Additional information provided on test reports may include the following: local percentile ranks, national norms, a breakdown of the various subject areas into subscales, the total number of items for each subscale, and the number of items answered correctly by the examinee. To help the company managers understand the report, an explanation of the information is usually provided. In addition, the company may share some or all of the results of the test with the applicant.

Stanines

The word stanine was derived from the words "**standard nine.**" **Stanines** comprise a scale with 9 points, a mean of 5, and a standard deviation of 2. In a bell shaped distribution, stanines allow the conversion of percentile ranks into nine larger units (see Figure 7.1). Thus, stanine 5 includes the middle 20% of the distribution; stanines 4 and 6, each includes 17%; stanines 3 and 7, each includes 12%; stanines 2 and 8, each includes 7%; and stanines 1 and 9, each includes 4% of the distribution. Approximately one-fourth of the scores in the distribution (23%, to be exact) are in stanines 1-3, and 23% of the scores are in stanines 7-9. Approximately one-half (54%) of the scores in the distribution is contained in stanines 4-6, the middle stanines.

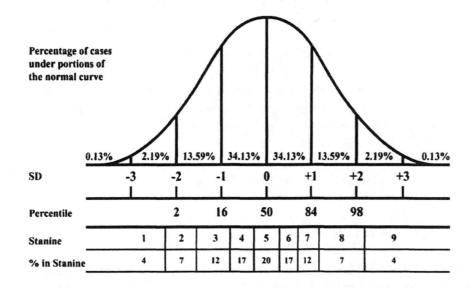

Figure 7.1   A graph showing standard deviations, percentiles, stanines, and percentages or scores in each stanine

Cut Scores

Another test interpretation tool used by managers is a cut score. **Cut scores** (or **cut off scores**) are established to divide the range of test scores into categories. These categories can be used to describe the test scores, or more commonly they are used to help managers make decisions about the examinees. For example, a minimum score on a pre-employment test may be required for a job applicant to be considered for a specific job. Professional certification and licensing exams require minimum scores that examinees must achieve to pass the exam and practice the profession.

There is no standard way of determining cut scores. Often cut scores are established based on the desired number of examinees that the person or organization administering the test wants to pass or not pass the test. For example, assume there are 1,000 people taking a pre-employment test for a company that has only 50 jobs available. Based on past experience with the test the company may set a high minimum score that examinees must achieve to be considered further for employment. A smaller number of applicants will achieve that score, thus reducing the number of people the company needs to consider for the 50 open jobs. In another case, if a company only has 60 applicants for 50 jobs, it may want to lower the cut score to ensure a larger enough number of applicants will survive the cut. Organizations need to be concerned about the potential legal implications of the arbitrary assigning of a cut score. In the case of determining a cut score for a licensing test, informed judgment must determine that examinees scoring less than the cut point are likely to make serious errors due to their lack of knowledge or skills. Obviously tests are not perfect. Some examinees who pass a test could still be incompetent and some who do not pass the test could have been successful practitioners in the profession. Testing companies monitor the relationship between their test results and the performance of those who take their tests to help organizations determine cut scores.

Criterion-Referenced Tests

Criterion-referenced (CR) tests are designed to compare the performance of an individual to certain criteria (unlike norm-referenced tests that compare the examinees to other people in the norming group).[1] The criteria, which should be specific and clear, are based on skills or objectives, as set forth by managers based on what they believe are the skills needed by employees for a particular job. After specifying the criteria, a task is designed to measure the level to which the criteria have been met. The task may be a pencil-and-paper achievement test, or it may be performance-based, such as simulated telephone conversations with clients.

Two main types of scores are used with CR tests: *percent correct* and *mastery/nonmastery*. Reporting scores in terms of **percent correct** is easy to do and to understand. These scores can be used to rank applicants for a job. An applicant that correctly answers 90% of the questions is ranked higher than one that successfully responds to 80% of the questions. This type of score does not take into consideration that the whole test may be too difficult or too easy. It is important that the test is a valid predictor of the examinees' ability to perform the job.

Reporting scores in terms of **mastery/nonmastery** is predicated on the theory of mastery learning that advocates mastery of the skill based on the completion of the test. There are several approaches that can be used to set the standards for mastery and to determine the point separating *mastery* from *nonmastery*. These different methods are similar to those discussed in the last section for determining cut points for passing/not passing pre-employment, professional certification or licensing tests.

While several published CR tests are available in specific areas, such as pre-employment and skills of current employees, many publishers also include CR interpretation in their NR tests. In addition to listing information about norms, such as percentiles and stanines, the computer-generated report sent to the organization for each examinee may also show the total number of items in each section of each subtest and the number of items answered correctly by the examinee. Similar information about others who have taken the test may also be included. This service can help the manager diagnose the specific strengths and weaknesses of individual applicants or employees.

---

1.   **A HINT:** Criterion-referenced tests may also be called *domain-referenced* or *content-referenced* tests.

Summary

1. There are several ways to report test scores. Some of the most common ways are: *raw scores, percent correct, standard scores* (such as *z scores*), *percentile ranks, stanines, cut scores,* and *percent correct.*

2. Raw scores obtained by an examinee on a manager-made test are usually converted into percent correct. Raw scores derived from standardized tests are usually converted into norms.

3. Tests can be classified into two major categories: *norm-referenced* and *criterion-referenced.* These two types differ in the way they are constructed and how they are used.

4. **Norm-referenced (NR)** tests include norms that allow the test user to compare the performance of an individual taking the test to that of similar examinees who have taken the test previously.

5. A **norming group** is a sufficiently large sample with demographic characteristics similar to those of potential test takers. Scores from the norming group are used to develop the test norms. When new examinees take the test, their scores are compared to the scores of the norming group.

6. Test items constructed for NR tests are written specifically to maximize differences among the examinees. Some items have a high level of difficulty in order to differentiate among the top examinees, while other items are easy in order to distinguish among the low-scoring examinees. Most items are of average level of difficulty and are designed to be answered correctly by 30-80% of the examinees.

7. *Technical manuals* of standardized tests include information about the test development process, the demographic characteristics of the norming sample, and other psychometric information (such as the test reliability and validity).

8. Testing companies may develop two types of norms: *national* and *local.* **National norms** compare the examinee to similar examinees in the population at large, while **local norms** compare the examinees to others with the same demographic characteristics or in the same organization.

9. A *percentile rank* describes the percentage of people who scored *at* or *below* a given raw score. Percentile ranks are easy to understand by lay people, which may be one of the reasons they are popular as norm-referenced scores.

10. A **percentile band** is often used to provide an estimated range of the true percentile rank. The bands are used due to the fact that tests are not completely reliable and include a certain level of error.

11. **Stanines** (derived from the words "**standard nine**") comprise a scale of norms that is used to convert percentile ranks into 9 larger units. The scale has a mean of 5, and a standard deviation of 2.

12. A **cut score** is a specified point or score on a test; scores at or above that point are interpreted or acted upon differently from scores below that point.

13. Cut scores are often used to determine who did/did not pass a test; the cut point can help managers decide who they want to consider hiring; they also determine who passes certification tests.

14. **Criterion-referenced (CR)** tests are designed to compare the performance of an individual to certain criteria. The criteria, which should be specific and clear, are based on skills or objectives, as set forth by managers. Two main types of scores are used with CR tests: *percent correct* and *mastery/nonmastery.*

15. Reporting scores in terms of **percent correct** is easy to do and to understand. It is often used by managers to rank employment applicants.

16. Reporting scores in terms of **mastery/nonmastery** is based on the theory of mastery learning that advocates mastery of the skill being tested before learning other skills.

Part Four

Measuring Relationships

Chapter 8

Correlation

The word correlation is used in everyday life to indicate a relationship or association between events or variables. However, in statistics, correlation refers specifically to the procedure used to quantify the relationship between two numerical variables through the use of a correlation coefficient.

Correlation may be defined as the relationship or association between two or more variables. These variables have to be related to each other or paired. There are several common ways to use correlation in the field of business, such as the correlation between the pre-employment test scores and performance on the job, or the correlations between advertising spending and sales revenue.

The strength or degree of correlation is indicated by a correlation coefficient. The coefficient can range from -1.00 indicating a perfect negative correlation; to 0.00, indicating no correlation; to +1.00, indicating a perfect positive correlation.

It is important to understand that correlation does not imply causation. Just because two variables correlate with each other does not mean that one caused the other. The only conclusion we can draw from a correlation between two variables is that they are related. In many cases, it is possible that a third variable causes both variables to correlate.

In addition to being used to describe the relationship between variables, correlation can also be used for prediction (in a statistical procedure called regression that is described in Chapter 9). Additionally, correlation can be used in assessing reliability (e.g., test-retest reliability; see Chapter 13) and in assessing validity (e.g., concurrent validity; see Chapter 14).

Graphing Correlation

Correlation between two measures obtained from the same group of people can be shown graphically through the use of a *scattergram*. A scattergram (or a scatterplot) is a graphic presentation of a correlation between two variables (see Figure 8.1). The two axes in the graph represent the two variables and the points represent pairs of scores. The *direction* of the correlation (positive or negative) and the *magnitude* of the correlation (ranging from -1.00 to +1.00) are depicted by a series of points. Each point is located *above* a person's score on the horizontal axis (the X variable) and *across* from that person's score on the vertical axis (the Y variable).

Notice that the points on the scattergram in Figure 8.1 create a pattern that goes from bottom left upward to top right. This is typical of a positive correlation, in which an *increase* in one variable is associated with an *increase* in the other variable. The points on the scattergram depicted in Figure 8.1 cluster together to form a tight diagonal pattern. This pattern is typical of a high (or very high) positive **correlation.**

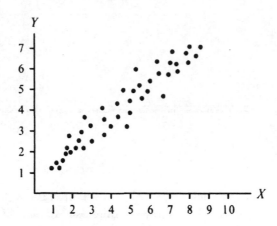

Figure 8.1   A scattergram showing a positive correlation between two variables, *X* and *Y*
(In thousands of dollars)

To illustrate the positive correlation displayed in the scattergram in Figure 8.1, let's look at the following fictitious study. In this study, the *X* variable is the level of advertising (measured in thousands of dollars spent by a company), and the *Y* variable is the same company's sales measured in thousands of dollars. According to Figure 8.1, an increase in a company's spending in advertising is associated with an increase in that company's sales. Conversely, companies that spend less on advertising have lower sales levels. In business, though, we rarely observe such high correlation between any two variables such as the advertising/sales example in Figure 8.1. It is also rare to see such high correlations of attitudes and behaviors of consumers or employees in business.

In a negative correlation, an *increase* in one variable is associated with a *decrease* in the other variable. For example, we can expect a negative correlation between the number of classes skipped by students during the semester and their grade point average (GPA). That is, as students cut classes for more and more days (an *increase* in *X*), their GPA falls lower and lower (a *decrease* in *Y*). The scattergram in Figure 8.2 shows the hypothetical relationship between the two variables. Note that in this negative correlation, the direction of the points is from top left downward toward bottom right.

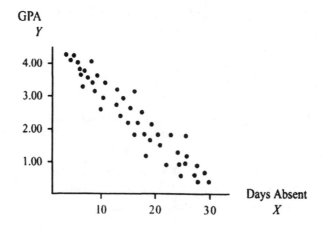

Figure 8.2   A scattergram showing a *negative* correlation between the number
of days students skip class and their grade point average (GPA)

If you were to draw an imaginary line around the points on a scattergram, you would notice that as the correlation (positive or negative) gets higher, the points tend to cluster closer and form a clear pattern (Figure 8.3). Thus, an inspection of the scattergram can indicate the approximate strength (or degree) of the correlation. For example, the scattergram in Part a where the points create a tight pattern shows a higher correlation than that in Part b where the points are spread out wider.

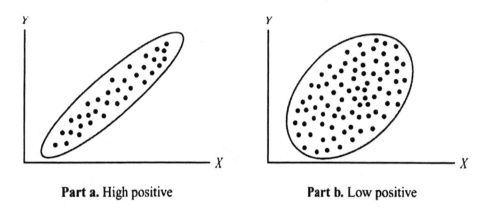

Part a. High positive   **Part b.** Low positive

Figure 8.3   Scattergrams depicting two positive correlations: A high positive
correlation (*Part a*) and a lower positive correlation (*Part b*)

When there is no correlation, or a very low correlation, between two variables, the scattergram contains points that do not form any clear pattern, and are scattered widely (see Figure 8.4).

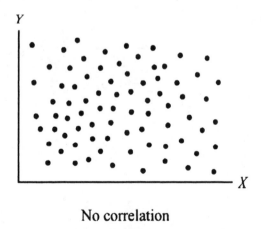

No correlation

Figure 8.4   A scattergram showing no correlation between two variables, *X* and *Y*

Scattergrams can also be used to locate a specific pair of scores. For example, let's examine Table 8.1 that lists the number of years an employee has worked for a company (variable X) and the employee's score on an employee satisfaction survey (variable Y) for 7 employees. Figure 8.5 depicts the data in Table 8.1.

Table 8.1   Years of Employment and Job Satisfaction
                   of Seven Employees

Employee Number	Years of Employment X	Job Satisfaction Score Y
A	18	20
B	17	15
C	11	12
D	19	18
E	13	12
F	15	16
G	17	18

Each point on the scattergram below represents one employee and corresponds to the scores listed in Table 8.1. For example, let's find the point located at the top right-hand side of the scattergram in Figure 8.5 that represents Employee G. We can draw an imaginary line from that point down toward the X axis (Years of Employment). Our line should intersect the axis at the score of 17. A horizontal line from point G toward the Y axis (Job Satisfaction Score) should intersect the axis at the score of 18. These scores—17 in Years of Employment and 18 in Job Satisfaction—are indeed the same as those listed for Employee G in Table 8.1.

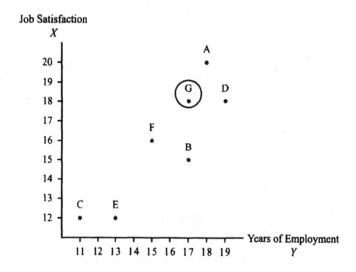

Figure 8.5   A scattergram of the correlation between Years of Employment and
               Job Satisfaction (data in Table 8.1)

A scattergram can help us identify scores that are noticeably different from the other scores. These scores, called outliers, can be easily spotted on a scattergram where they fall outside the range and pattern of the other points.[1]

---

1.   **A HINT**: Outliers are not unique to correlation. There may be outliers in any distribution caused by various reasons. Researchers may wish to pay special attention to outliers and study them further.

Pearson Product Moment

The most commonly used correlation procedure is the Pearson product moment. Pearson product-moment coefficient (often referred to as Pearson *r*) is named in honor of Karl Pearson (1857-1936), a British scientist who contributed a great deal to the development of statistics. Pearson was a student of Sir Francis Galton who studied heredity. In 1896, Pearson developed the product-moment coefficient, which became quite popular within a short period of time. In order to use Pearson's correlation, the following requirements should be satisfied:

1. The scores are measured on an *interval* or *ratio* scale.

2. The two variables to be correlated should have a *linear* relationship (as opposed to *curvilinear* relationship). To illustrate the difference between linear and curvilinear relationships, examine Figure 8.6. *Part a* shows a *linear* relationship between height and weight, where the points form a pattern going in one direction. *Part b* shows a *curvilinear* relationship where the age of individuals is correlated with their strength. Notice that the direction of the points is not consistent. In this example, the trend starts as a positive correlation and ends up as a negative correlation. For example, newborns are very weak, and get stronger with age. They then reach an age when they are the strongest, and as they age, they become weaker. When Pearson *r* is used with variables that have a curvilinear relationship, the resulting correlation is an *underestimate* of the true relationship between these variables.

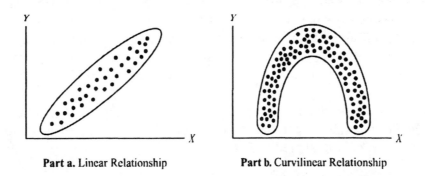

Part a. Linear Relationship   Part b. Curvilinear Relationship

Figure 8.6   Scattergrams showing *linear* relationship between height and weight
(*Part a*), and *curvilinear* relationship between age and strength (*Part b*)

When observations on the variables to be correlated are rank-ordered, the statistic known as Spearman rank-order correlation is used and the correlation coefficient is represented by r_s.[2] This rank-order correlation coefficient is interpreted in the same way as the Pearson coefficient *r*.

Interpreting the Correlation Coefficient

After obtaining the correlation coefficient, the next step is to evaluate and interpret it. It is important to remember that the *sign* of the correlation (negative or positive) is not indicative of the *strength* of the correlation. A negative correlation is not something negative. What matters is the *absolute value* of the correlation. Thus, a negative correlation of -.93 indicates a stronger relationship than a positive correlation of +.85.

---

2.   **A HINT**: The Greek letter r (rho) may also be used to indicate the rank-order coefficient.

Table 8.2 lists guidelines for the interpretation of the strength of the correlation coefficients. These guidelines apply to both positive and negative correlation coefficients. There is no clear consensus among researchers as to the exact definition of each category and the categories listed in the table overlap. Thus, it is the researcher's decision how to define a certain coefficient. For example, one researcher may describe a correlation coefficient of .68 as high; another researcher may define it as moderate; and a correlation of .40 may be described as low or moderate. You can also use two categories to define a coefficient. For example, you can describe a correlation of .40 as low-to-moderate and a correlation of .60 can be described as moderate-to-high.

Table 8.2	An Interpretation of Correlation Coefficients
Correlation	Interpretation
.00 - .20	Negligible to low (a correlation of .00 would be defined as "no correlation")
.20 - .40	Low
.40 - .60	Moderate
.60 - .80	High/substantial
.80 - 1.00	Very high (a correlation of 1.00 would be defined as a "perfect correlation")

Another way to evaluate the correlation coefficients is to divide the coefficients between .00 and 1.00 into three categories. Coefficients between .00 and .35 would be defined as low; coefficients between .35 and .66 would be considered moderate; and coefficients between .67 and 1.00 would be considered high. Again, two categories can be used to describe borderline coefficients. For example, coefficients such as .28 or .36 can be described as low-to-moderate; and coefficients such as .60 or .68 can be described as moderate-to-high.

In describing and assessing correlations, it is important to consider the purpose of the study and the potential use of the results of the study. For example, a correlation coefficient of .60 may be adequate for the purpose of group prediction but insufficient for individual prediction purposes.

The statistical significance level (p value) is often reported along with the coefficient itself. However, if the study involves the whole population and there is no attempt to generalize the results to other groups or settings, then the p value is not of importance. Rather, the obtained correlation coefficient is used to indicate the relationship between the variables.

The level of statistical significance is greatly affected by the sample size and might imply a high level of association between variables even when the correlation is low. For example, with a large sample size of 300, even low correlations such as $r = .12$ are reported to be significant at $p < .05$ and a correlation of .15 is reported as significant at the $p < .01$ level. Therefore, it is always a good idea to consider the practical significance of the correlation, along with its statistical significance.

Hypotheses for Correlation

The null hypothesis (H_0) states that in the population the correlation coefficient expressing the relationship between the two variables being studied is zero:

$$H_O : r = 0$$

The alternative hypothesis (H_A) states that the population correlation is *not equal* to zero:

$$H_A : r \neq 0$$

After we obtain the correlation coefficient, we then consult the table of critical *(r)* values for Pearson correlation coefficients. To use the table, we have to calculate the degrees of freedom *(df)* for the study. In correlation, the degrees of freedom are the number of *pairs* of scores minus 2. If the *obtained* correlation coefficient *(r) exceeds* the critical value, the null hypothesis is *rejected*. Rejecting the null hypothesis means that the chance that the correlation coefficient is 0 *(r=0)* is very small, and that *r* is large enough to be considered different from zero. When the obtained coefficient is *smaller* than the critical value, the null hypothesis is *retained*. We conclude that there is a high degree of likelihood that the correlation is not significantly different from 0.

When the null hypothesis is rejected, the level of significance (*p* level) is reported. This can be done using two approaches. There are researchers who choose to use the conventional "benchmarks" approach where the level of significance (*p* level) is listed usually as $p<.05$, $p<.02$, $p<.01$ or $p<.001$. Other researchers prefer to report the *exact* level of significance. Computer statistical packages usually print the exact *p* values thus making this information readily available to researchers.

Computing Pearson Correlation

The correlation coefficient is reflective of the relative position of scores in their group. The correlation will be *high* and *positive* if the following occurs: high scores on *X* are also high scores on *Y*; low scores on *X* are also low scores on *Y*; and scores in the middle on *X* are also scores in the middle on *Y*. The actual scores on *X* and on *Y* do not have to be the same, only the relative position of scores in their group. The correlation will be *high* and *negative* if high scores on one variable are associated with low scores on the other variable and low scores on the first variable are associated with high scores in the other variable. When there is no pattern of relationship between the scores on the two variables, the correlation will be low positive or low to very low positive, or low to very low negative.

To illustrate the interpretation of Pearson correlation, suppose we want to find the correlation of the results of six salespeople's performance based on Number of Sales Calls Closed (*X*) and Number of Cold Calls Attempted (*Y*) (Table 8.3). The correlation between the two variables (*X* and *Y*) is *r*=.95.

Table 8.3   Sales Calls Closed (*X*) and Cold Calls Attempted (*Y*) for Six Salespeople

	Sales Calls Closed	Cold Calls Attempted
Salesperson	X	Y
A	9	14
B	8	12
C	10	14
D	11	15
E	10	15
F	7	11

After the correlation coefficient is calculated, the next step is to ascertain whether it is statistically significant. Table 8.4 shows a part of the table of *critical values* of the Pearson *r* coefficient. To use the table, we first need to determine the degrees of freedom (*df*). In correlation, the degrees of freedom are the number of pairs of scores minus two. In our example, we have 6 salespeople; therefore, the degrees of freedom (*df*) are 4 (the number of pairs minus 2). The critical values for 4 degrees of freedom are .811 for a *p* level of .05, .882 for a *p* level of .02, and .917 for a *p* level of .01. These critical values can be listed as: $r_{(.05,4)} = .811$, $r_{(.02,4)} = .882$, and $r_{(.01,4)} = .917$. The .05, .02, and .01 listed inside the parentheses indicate the *p* level, and the number 4 indicates the degrees of freedom. Our calculated *r* value of .95 *exceeds* the critical values listed for .05, .02, and .01 levels of significance. Consequently, we *reject* the null hypothesis at *p* <.01 level. We conclude that a correlation coefficient of this magnitude (*r*=.95) could have occurred by chance alone less that 1 time in 100. (See Chapter 2 for a discussion of the statistical hypothesis testing.)

Table 8.4   Sample Values of the Correlation Coefficient (Pearson's *r*) for Different Levels of Significance

df	*p* level .10	*p* level .05	*p* level .02	*p* level .01
3	.805	.878	.934	.959
4	.729	.811	.882	.917
5	.669	.754	.833	.874

Factors Affecting the Correlation

Besides the relationship between the relative positive scores on the two variables, there are other factors that affect the correlation. The *reliability* of the instruments used to collect data may affect the correlation. The correlation coefficient may *underestimate* the true relationship between two variables if the measurement instruments used to obtain the scores are not reliable. (See Chapter 13 for a discussion of reliability.)

The correlation obtained may also underestimate the real relationship between the variables if one or both variables have a *restricted range* (i.e., low variance). To demonstrate an extreme case, suppose all the students in the course

Principles of Statistics receive the same score on test *X*. (This may happen if a test is too easy.) If we try to correlate their scores on *X* with their scores on another test, *Y*, we will get a correlation of zero (*r*=.00).

To illustrate this point with a numerical example, let's look at the scores of four students on two tests, *X* and *Y*, which are listed in Table 8.5. Notice that all the students received the same score on *X* (*X*=25). Therefore, there is no variability on that variable and the variance of the *X* scores is zero.

Table 8.5	Scores of Four Students on Two Tests	

Student	Test *X*	Test *Y*
A	25	15
B	25	18
C	25	17
D	25	14

The scattergram showing this correlation can be found in Figure 8.7. Note that the points that represent the students form a vertical line because all the scores on the *X* variable have the same value, even though the same four students have different scores on Y. Knowing the student's score on the *X* variable will not allow us to predict the student's score on the *Y* variable.

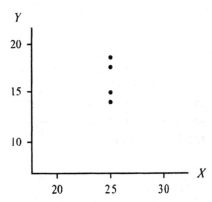

Figure 8.7   A scattergram of the data in Table 8.4 showing *r* = .00, when all the scores on the X variable are the same

The Coefficient of Determination and Effect Size

When interpreting the correlation coefficient, researchers describe it in terms of its magnitude (e.g., low, moderate, or high) and its level of statistical significance (*p* value). Another index, called the *coefficient of determination* or *shared variance*, may also be used to evaluate the correlation. The coefficient of determination describes how much individual differences in one variable are associated with individual differences in the other variable. The coefficient of determination can be thought of as the percentage of the variability in one variable that can be attributed to differences in the scores on the other variable. This index is often used in prediction studies where one variable is used to predict another. (See Chapter 9, which discusses prediction and regression.) The coefficient of determination is computed by squaring the correlation coefficient *r*.

For example, suppose a college is using an entrance test that is given to all students who apply for admission. Assume, further, that based on results from past years, the admission office in that college computes the correlation between past students' admission test scores and their college GPA and find the correlation to be .50 (r=.50). Calculating the coefficient of determination would allow the admission office to assess the proportion of variability in one variable (college GPA) that can be explained or determined by the other variable (admission test score). The coefficient of determination is the square of the correlation coefficient. In our example, with a correlation coefficient of .50, the coefficient of determination is .25, or 25% (r^2=.50^2=.25%).

Figure 8.8 shows the overlap between the two variables, the admission test and college GPA. As we can see, the variables overlap 25%. Knowing the score of a person on one variable (admission test), one can predict 25% of this person's performance on the other variable (college GPA). Three-fourths (75%) of their performance are still unknown and unaccounted for by the predictor variable (admission test).

25%

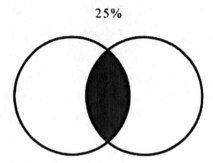

Figure 8.8   A graphic presentation of the relationship between two variables, Admission Test and College GPA when r=.50 and the *coefficient of* determination (r^2) is 25%

When there is a high correlation between two variables, there is a greater overlap between the variables. For example, when the correlation between two variables is .90, the coefficient of determination is 81% (r^2=.90^2=0.81=81%). Figure 8.9 illustrates the overlapping of the two variables.

81%

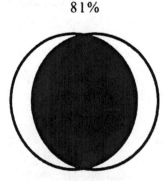

Figure 8.9   A graphic presentation of the relationship between two variables when r=.90 and the *coefficient of determination* (r^2) is 81%

As was discussed in Chapter 2, the correlation coefficient may be used to compute an *effect size*. The effect size can be used to evaluate the practical significance of the obtained correlation. Most researchers use the coefficient of determination (r^2) as an index of effect size, although others suggest using the correlation coefficient itself (r). In our example of the correlation between the college admission test and the students' GPA, an effect size of r=.50 (r^2= .25) would be considered large. However, keep in mind that only 25% of the variance in the dependent variable (college GPA) is predicted by the admission test.

Intercorrelation Tables

At times, researchers are interested in correlating several variables with each other. Instead of embedding the results from such analyses in the text narrative, the results can be presented in an intercorrelation table. An intercorrelation table is used to display the correlations of several variables with each other. For example, let's say a real estate broker wants to take an in-depth look at the market price of homes by determining if there are correlations between five factors she believes affect the price. These factors are: Size of the Home (in square feet), Number of Bedrooms, School District Rating, and Lot Size. Prices for selected homes were also included in the analysis. The real estate broker displays the results in an intercorrelation table (see Table 8.6).

Table 8.6   Intercorrelations of Five Factors

	1	2	3	4	5
	Size	Bedrooms	District	Lot	Price
1. Size	1.00	.63	.45	.57	.82
2. # of Bedrooms	.63	1.00	.39	.68	.78
3. School District	.45	.39	1.00	.85	.86
4. Lot Size	.57	.68	.85	1.00	.91
5. House Price	.82	.78	.86	.91	1.00

As you can see, Table 8.6 has two distinct features:

1. All the correlations that are listed on the diagonal line in the center of the table (from top left to bottom right) are perfect (r=1.00). The reason is obvious: these are all correlations of a variable with itself (e.g., Size, number of Bedrooms, School District Rating, Lot Size, House Price).

2. If we divide the table into two triangles: top right and bottom left, we can see that the correlations coefficients that are recorded in the two triangles are the same and are a mirror image of each other.

Considering these two features, it is clear that Table 8.6 contains *duplicate* information and *unnecessary* information (the correlations of 1.00 on the diagonal). Thus, the table can be reorganized to present the results more efficiently. Table 8.7 may look as if it lacks some information, but, in fact, it contains all the information needed.

Table 8.7	A Revised Intercorrelation Table (Same Data as in Table 8.6)			
	2	3	4	5
	Bedrooms	District	Lot	Price
1. Size	.63	.45	.57	.82
2. # of Bedrooms		.39	.68	.78
3. School District Rating			.85	.86
4. Lot Size				.91
5. House Price				

You may find published reports where the lower left-side triangle is used to display the information, instead of the top right-hand side, as is the case in Table 8.7. You may also find intercorrelation tables that include additional information about the measures that are being intercorrelated. For example, some of these tables may also list the means and standard deviations of the measures.

In the intercorrelation tables there are two ways to identify those correlation coefficients that are statistically significant. One approach is to display the significance levels with asterisks that indicate certain levels of significance (e.g.,* $p < .05$ or ** $p < .01$). The other approach is to list the *exact* level of significance next to each correlation coefficient (e.g., $p = .046$, or $p = .003$).

Correlation Tables

Correlation tables differ from *intercorrelation* tables in the type of information they convey and in their layout. Both types of tables provide an efficient way to present a large number of correlation coefficients. In correlation tables, the variables listed in the rows and columns are different from each other.

To illustrate the use of a correlation table, suppose we want to examine the job performance levels of 216 subordinates and their managers. The subordinates in the study are divided into three groups, those with low, medium, and high performance levels. The gender of each subordinate is recorded in addition to the type of manager they have democratic or autocratic (Table 8.8).[3] Note that in this example we do not have two scores for each participant as was the case with other examples in this chapter. Instead, in this example each subordinate is paired with his/her manager.

---

3.   **A HINT**: As with any continuous variable that is divided into categories (such as high, medium, and low), the criterion used for creating the categories has to be logical.

Table 8.8   Correlations of Democratic and Autocratic Managers
with Their Male and Female Subordinates with
Different Performance Levels

	Low Level Performers		Average Level Performers		High Level Performers	
	Female	Male	Female	Male	Female	Male
Democratic Managers	r=.56* n= 15	r=.62 n= 20	r=.55** n= 22	r=.44* n= 21	r=.48 n= 16	r=.58* n= 15
Autocratic Managers	r=.52 n= 14	r=.45 n= 18	r=.54** n= 23	r=.48* n= 20	r=.49* n= 17	r=.50 n= 15

* p <.05   ** p <.01

In addition to the correlation coefficient (r) in Table 8.8, the level of significance and the sample size (n) are also recorded in each cell. For example, according to this table, we have correlational data for 15 females and their democratic managers (n=15) in the low performance category. The correlation of the scores of these female employees and their democratic managers is .56 (r=.56), significant at the .05 level (*p<.05).

Summary

1. **Correlation** is defined as the relationship or association between two or more paired variables. The most common way to pair variables is to administer two measures to the same group of people and correlate their scores on the two measures.

2. The **correlation coefficient** indicates the strength (or degree) of correlation. The coefficient can range from +1.00 (perfect positive correlation) to -1.00 (perfect negative correlation). A coefficient of 0.00 indicates no correlation.

3. Correlation does not imply *causation*. Just because two variables correlate with each other does not mean that one caused the other.

4. Correlation is used to describe relationships between variables in prediction studies and in the assessment of reliability and validity.

5. A **scattergram** (or a **scatterplot**) is a graphic presentation of a correlation between two variables. The two axes in the graph represent the two variables and the points represent pairs of scores.

6. The *direction* of the points on the scattergram and the degree to which they *cluster* indicate the strength of the correlation and whether the correlation is positive or negative. A scattergram can also show whether there are scores that are outliers.

7. In a **positive** correlation, an *increase* in one variable is associated with an *increase* in the other variable. In a **negative** correlation, an *increase* in one variable is associated with a *decrease* in the other variable.

8. The most commonly used correlation procedure is the **Pearson product moment**, whose coefficient is represented by the letter r. Pearson r is used with data measured on an interval or a ratio scale when the variables to be correlated have linear relationship.

9. Correlation coefficients can be described using words such as negligible, low, moderate, high, and very high. A combination of categories may also be used, such as moderate to high.

10. In describing and assessing correlations, it is important to consider the purpose of the study and the potential use of the results of the study. It is also important to consider the *practical* significance of the correlation and the *effect size*, along with its *statistical* significance.

11. The *null hypothesis* in correlation states that in the population the correlation coefficient is zero, and the *alternative hypothesis* states that in the population the correlation is not equal to zero.

12. The obtained correlation coefficient may be an *underestimate* of the real relationship between the variables if one or both variables have a low reliability or if one or both variables have a *restricted range* (i.e., low variance).

13. The **coefficient of determination** (or **shared variance**) describes how much individual differences in one variable are associated with individual differences in the other variable. This index is often used in prediction studies where one variable is used to predict another. The coefficient of determination is found by squaring the correlation coefficient. It can also be used as the *effect size* to assess the practical significance of the study's results.

14. Correlations between three or more variables are often presented in an **intercorrelation** or **correlation** table.

Chapter 9

Prediction and Regression

In our daily life, prediction is quite common. When we hear thunder and see lightning, we often predict they will be followed by rain. We also might predict the relationship between the day of the week and the expected crowd at the movie theater. In business, we also use prediction. For example, we might predict that a bright college graduate will do well in her first job, or that an employee who is having difficulties learning new job skills is probably going to get a poor evaluation.

From our personal experience we know that our predictions do not always materialize, and people and events continue to surprise us. Sometimes rain does not follow thunder and lightning, and occasionally, bright college graduates do poorly in their first jobs. Nevertheless, knowing something about the relationship between the variables allows us to make a prediction that is better than a chance guessing.

Prediction is based on the assumption that when two variables are correlated, we can use one of them to predict the other. The discussion in this chapter focuses on using prediction in business settings.

The variable used as a predictor is the **independent variable** and it is represented by the letter X. The predicted variable, represented by the letter Y, is called the **criterion variable**, or the **dependent variable**. For example, the Graduate Management Admission Test (GMAT) may be used as a predictor variable, and graduate business school grade point average (GPA) may be the criterion variable.

The technique used for prediction is called **regression**. When only one variable is used to predict another, the procedure is called **simple regression**, and when two or more variables are used as predictors, the procedure is called **multiple regression**.

The discussion that follows focuses on simple **linear regression**, where the predictor variable (X) and the criterion variable (Y) have a linear relationship.[1] Our numerical example demonstrates the computations involved in simple regression. Additionally, the concept of *multiple regression* will be introduced briefly, without the use of a numerical example.

Simple Regression

After observing a high correlation between two variables, a researcher may want to use one variable to predict the other one. For example, suppose a business school dean notices that MBA students with higher GMAT scores are better students. The dean may want to conduct a study to explore the idea of using GMAT scores of incoming MBA students to predict their GPAs. The dean examines and analyzes GMAT scores and GPAs of a random sample of MBA students. If GMAT scores of MBA students are shown to be a good predictor of graduate GPAs, the dean may want to use this information to help him choose which students he wants to admit to the MBA program.

---

1. **A HINT**: See Chapter 8 for a discussion of the concept of linear relationship in correlation.

Because correlation does not imply causation, we cannot conclude from the regression study that the students' GMAT scores have an effect on their GPAs. Quite likely, both variables are related to aptitude, IQ or motivation. To ascertain whether GMAT scores affect GPAs, an *experimental* study would need to be conducted.

The prediction of scores of a group of people on one variable from their scores on another variable can be done by using a **regression equation**. The equation is used to draw a line that is used for prediction. In order to develop the equation, we first need to have the predictor and criterion scores for a group of people. The members of the group should be *similar* to those whose criterion scores we would like later to be able to predict. Once the equation is available, it can be used to predict criterion (dependent) scores for a new group of people for whom only the predictor scores are available.

The regression equation can be used to draw a line through a scattergram of the two variables involved, designated as *X* and *Y*. This line is called the **regression line**, or the **line of best fit** (Figure 9.1).

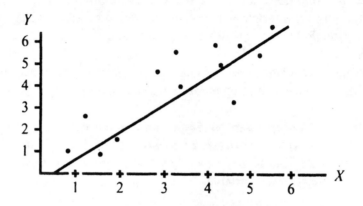

Figure 9.1   A scattergram showing the regression line

The position of the line is determined by the **slope** (the angle) and the **intercept** (the point where the line intersects the vertical axis, *Y*). The slope is represented by the letter *b* and the intercept is represented by the letter *a*. The slope may also be referred to as the **coefficient** and the intercept may be referred to as the **constant**. The prediction equation is:

$$Y' = bX + a$$

Where *Y'*	=	Predicted *Y* score
b	=	Slope
X	=	Score on the independent variable
a	=	Intercept

If you were to inspect regression lines and their slopes (*b*), you would realize that the higher the value of *b*, the steeper the line; and the lower the value of *b*, the flatter the line. Figure 9.2 shows four regression lines when the intercept is zero (i.e., the regression line passes through the point where both axes are at zero). When *b*=0.25, for every increase of 1 unit in *X*, there is an increase of 0.25 unit in *Y* (*Part a*); and when *b*=0.5, for every increase of 1 unit in *X*, there is an increase of 0.5 units in *Y* (*Part b*). When *b*=1, for every increase of 1 unit in *X*, there is an increase of 1 unit in *Y* (*Part c*); and when *b*=2, for every increase of 1 unit in *X*, there is an increase of 2 units in *Y* (*Part d*).

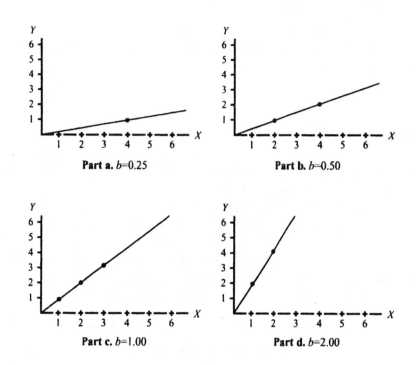

Part a. b=0.25

Part b. b=0.50

Part c. b=1.00

Part d. b=2.00

Figure 9.2   Four regression lines with different slopes

The Standard Error of Estimate (S_E)

Unless the predictor and the criterion variables have a perfect correlation, any attempt to use X (the predictor) to predict Y' (the criterion) is likely to result in a certain degree of error. Consequently, for some people the Y' score is an *overestimate* of their "true" Y score, while for others, the Y' score is an *underestimate* of their "true" Y score. The difference between the *actual Y* score and the *predicted Y* score for each individual is called the **error score** (or **residual**). The standard deviation of the error scores, across all individuals, is called the **standard error of estimate (S_E).**[2] More specifically, assuming a normal distribution of the error scores, the actual Y score would lie within $\pm 1 S_E$ of the Y' score about 68% of the time, and within $\pm 2 S_E$ about 95% of the time. The S_E is calculated by using the scores from the group used to generate the regression equation. The formula for S_E is:

$$S_E = S_Y \sqrt{1 - r^2}$$

Where S_E =   Standard error of estimate
S_Y =   Standard deviation of the Y variable
r^2 =   Square of the correlation [3]

Holding S_Y constant, S_E *decreases* as r *increases*. Thus, the *higher* the correlation, the *lower* the S_E, and when the standard error of estimate is lower, the prediction is more accurate. The following formula demonstrates that when X and Y have a perfect correlation ($r=1.00$), there is no error in prediction and the standard error of estimate (S_E) is zero:

$$S_E = S_Y \sqrt{1 - 1.00^2} = S_Y \sqrt{1 - 1.00} = S_Y \sqrt{0.00} = 0.00$$

2.   **A HINT:** The standard error of estimate may also be represented by the symbol Sy•x.
3.   **A HINT:** You may recall that in Chapter 8 we discussed the concept of r^2 (the coefficient of determination or shared variance) that refers to the proportion of the variability (or information) of Y that is contained in X.

When $r = 0.00$ (there is no correlation), S_E is equal to the *SD* of the *Y* variable (S_Y):

$$S_E = S_Y \sqrt{1 - 0.00} = S_Y \sqrt{1} = S_Y(1) = S_Y$$

An Example of Simple Regression

Ms. Wright, a manager in a ball bearing factory is concerned that older machines in the plant are creating a larger number of defective ball bearings than the newer machines. Her boss wants her to buy some more used machines but she believes the company should buy new ones. The manager hypothesizes that the age of a machine (in months) is a good predictor of the number of defective ball bearings. To ascertain whether the age of the machines is a good predictor of defective ball bearings, the manager uses the number of months a machine has been in operations in the plant (the *predictor*, or *independent* variable) and the number of defective ball bearings (the *criterion*, or *dependent* variable) from last year's production report to generate the regression equation. In this computational example, the age and defect rates of 10 randomly selected machines are used to demonstrate how to generate the regression equation (Table 9.1).

Table 9.1   **Age (in Months) and Number of Defects for**
                         10 Ball Bearing Machines

Machine	Age (Months) X	Defects Y
A	45	40
B	45	46
C	46	37
D	50	49
E	35	31
F	47	50
G	23	32
H	46	48
I	40	44
J	41	39
Mean	$\bar{X} = 41.80$	$\bar{Y} = 41.60$
Standard Deviation	$S_X = 7.84$	$S_Y = 6.88$

The manager finds that the correlation between the age of the machines and the number of defective parts is $r = .76$.[4] Next, the manager computes the *b* coefficient, followed by the computation of the value of *a*.

---

4.   **A HINT:** The computations of the correlation coefficient are not included here. (See Chapter 8 for a discussion of correlation).

$$b = r\frac{S_Y}{S_X} = (.76)\frac{6.88}{7.84} = (.76)(.88) = 0.67$$

$$a = \overline{Y} - b\overline{X} = 41.60 - (0.67)(41.80) = 41.60 - 28.03 = 13.57$$

After finding the values of b (the slope) and a (the intercept), they can be entered into the regression equation.

$$Y' = 0.67(30) + 13.57 = 20.1 + 13.57 = 33.67$$

Of course, using this equation to predict the number of defective units produced by a used machine the company is considering buying is predicated on the assumption that the used machines being considered for purchase are similar to the machines that were used to derive the regression equation.

Computing the Standard Error of Estimate

After creating the regression equation, the manager computes the *predicted* defects (the criterion score Y') for the 10 machines. Table 9.2 lists the age of each machine (X), the *actual* number of defects for each machine (Y), as well as the *predicted* number of defects for each machine (Y'). To compare the Y' defects (the predicted defects) to the actual Y defects, the manager creates a fourth column where the manager records the difference between the two Y scores (obtained Y defects minus Y' defects). This column is titled the *Error Score*. Note that for some of the machines, the predicted defects are an *overestimate*, resulting in a *negative* error score; for others, the predicted score is an *underestimate*, resulting in a *positive* error score. Checking the computation, you can see that the sum of the Error Score column is zero.

Table 9.2   Obtained Defects, Predicted Defects, and Error Defects of 10 Machines

Machine	X	Y	Predicted Defects Y'	Error Score E=Y-Y'
A	45	40	43.75	-3.75
B	45	46	43.75	2.25
C	46	37	44.42	-7.42
D	50	49	47.10	1.92
E	35	31	37.04	-6.04
F	47	50	45.09	4.91
G	23	32	28.99	3.01
H	46	48	44.42	3.58
I	40	44	40.39	-3.61
J	41	39	41.06	-2.06

$$\Sigma E = 0.00$$

Next, the manager computes the standard error of estimate (S_E):

$$S_E = S_Y \sqrt{1 - r^2} = 6.88\sqrt{1 - .76^2} = 6.88\sqrt{1 - .58} =$$
$$= 6.88\sqrt{.42} = 6.88(.65) = 4.44$$

Graphing the Regression Equation

As was mentioned before, the slope and the intercept in the regression equation can be used to draw a line through a scattergram that depicts the correlation of the two variables. Figure 9.3 shows a scattergram of the actual age obtained of the machines (X) and the number of defective ball bearings produced (Y), with the regression line added. All the predicted Y' scores lie on the regression line.[5] Note that the regression line goes through the intersection of the means of the two variables (the mean of X is 41.80 and the mean of Y is 41.60).

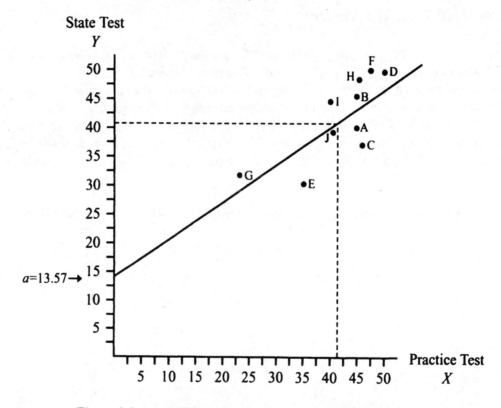

Figure 9.3   A regression line for the predicting the number of defective ball bearings of 10 machines using the age of the machines as a predictor (see data in Table 9.2)

As we can see, for several machines, the Y' score is an overestimate (i.e., the Y' score is above the actual Y score); and for several machines, the Y' score is an underestimate (i.e., the Y' score is below the actual Y score).

---

5.   **A HINT**: Computer statistical programs can be used to draw the line.

The Coefficient of Determination (r^2)

The *coefficient of determination* (r^2) can be used to describe the relationship between the variables.[6] In our example, the manager used the age of the machines in the factory to predict the number of defective units that would be produced. The manager found that the correlation between the two variables was $r=.76$. As you may recall, to find the coefficient of determination we need to square the correlation (r^2). With $r=.76$, the coefficient of determination is $.76^2 = .5897$ (or 59%). This coefficient means that about 59% of the variation in the number of defective ball bearings (Y) can be accounted for by the age of the machines (X); 41% of the variation is due to other factors.

Multiple Regression

The **multiple regression** procedure is employed when two or more variables are used to predict one criterion variable. For example, the amount a company spends on advertising, combined with the number of follow-up calls salespeople make, may be used to predict a company's revenue for the month. A researcher is likely to consider using several variables as predictors when there is no single variable that has a high correlation with the criterion, so as to serve as a satisfactory single predictor. In our example, combining the two predictor variables (the amount of money spent on advertising and the number of follow-up calls) is likely to predict the criterion variable more accurately, compared with using only one of the two predictors.

The regression equation in multiple regression is an extension of the equation for simple regression. In addition to the intercept (a), the equation contains a regression coefficient (b) for each of the predictor variables (X). The combined correlation of the predictor variables with the criterion variable is called **multiple correlation**, represented by the symbol R. With two predictors, the equation is:

$$Y' = b_1 X_1 + b_2 X_2 + a$$

Where Y'	=	Predicted Y score
b_1	=	Slope (coefficient) of predictor X_1
X_1	=	Score on independent variable (predictor) X_1
b_2	=	Slope (coefficient) of predictor X_2
X_2	=	Score on independent variable (predictor) X_2
a	=	Intercept (constant)

In multiple regression, the *coefficient of determination* is represented by R^2 which is similar to r^2 in simple regression. Just like r^2, the coefficient of determination in multiple regression can range from 0 to 1.00. R^2 indicates the proportion of the variation in Y that can be accounted for by the variation of the *combined* predictor variables.

For any level of correlation between the predictor variables and the criterion, when the predictor variables have a low correlation with each other, R^2 is greater than when the predictor variables correlate highly with each other. To illustrate this point, let's look at Figure 9.4. *Part a* shows two predictor variables, X_1 and X_2, which correlate highly with the criterion variable Y. In addition, the two predictors also correlate highly with each other. The high correlation between the two predictor variables is evidenced by the fact that they overlap a great deal. Adding a second predictor (X_2) to the first predictor (X_1) has not significantly increased R^2, the amount of variation in Y (the criterion) that can be accounted for by the predictors. That is, adding a second predictor does not account for a much greater

---

6.   **A HINT:** This concept was discussed in Chapter 8 (Correlation).

proportion of the criterion *Y*. As we can tell, in *Part a* there is a large white area on the left side that represents the proportion in variable *Y* that is not accounted for by the two predictors.

Part b shows two predictor variables, X_3 and X_4, and a criterion variable *Y*. Each of the two predictors has a *high* correlation with the criterion variable, and a *low* correlation with each other (X_3 and X_4 overlap very little). In *Part b*, the small white area in variable *Y* shows that the two combined predictor variables cover more of the criterion variable and can account for more of the variation in *Y*.

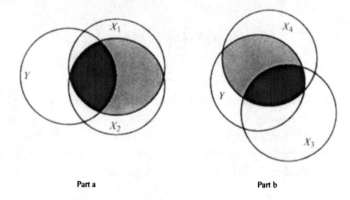

Part a   Part b

Figure 9.4   Two graphs showing different levels of correlation between the two predictors, and between the predictors and the criterion: the two predictors, X_1 and X_2, correlate highly with each other (*Part a*); and the two predictors, X_3 and X_4, have a low correlation with each other (*Part b*)

Summary

1. **Regression** is a technique used for prediction. The variable used as a predictor is called **the independent variable** and it is represented by the letter *X*. The predicted variable is called the **criterion** or the **dependent variable**, and is represented by the letter *Y*.

2. Regression is based on the assumption that the predictor (or predictors) and the criterion variable correlate with each other. The higher the correlation, the more accurate the prediction.

3. When one variable is used to predict another variable, the procedure is called **simple regression**. When two or more variables are used as predictors, the procedure is called **multiple regression**.

4. **Linear regression** is used when the predictor variable (*X*) and the criterion variable (*Y*) have a linear relationship.

5. The **regression equation** is used to predict *Y* scores for a given group of individuals for whom the *X* scores are available. The predicted *Y* score is represented by *Y'*. The equation is:

    $$Y' = bX + a$$

6. The regression equation can be used to draw a line. The position of the line is determined by its **slope**, represented by the letter *b*; and by its **intercept** (the point where the regression line intersects the vertical axis), represented by the letter *a*. The slope (*b*) may also be identified as the **coefficient**, and the intercept may be identified as the **constant**.

7. The characteristics of the sample used to derive the regression equation should be similar to those of the sample for which we want to predict future scores.

8. When inspecting regression lines, we can see that the higher the value of b, the steeper the line; and the lower the value of b, the flatter the line.

9. Unless the predictor and the criterion variables have a perfect correlation, any attempt to use the predictor variable (X) to predict the criterion variable (Y') is likely to result in a certain degree of error. The **error score** (or the **residual**) is the difference between actual Y and Y' scores.

10. The standard deviation of the error scores across all individuals is called the **standard error of estimate (S_E)**. S_E indicates how much, on the average, Y' scores *overestimate* or *underestimate* the actual Y scores.

11. As the correlation between variables increases, the S_E decreases thereby making the prediction more accurate.

12. The *coefficient of determination*, r^2, can be used to describe the relationship between the two variables. It describes the amount of variation in the scores on the criterion variable Y that can be accounted for by individual differences on X, the predictor variable. (r^2 was discussed in Chapter 8.)

13. The **multiple regression** procedure is employed when two or more variables are used to predict one criterion variable. A researcher is likely to consider using two or more variables as predictors when there is no single variable that has a high correlation with the criterion, so as to serve as a satisfactory predictor. In such cases, additional predictor variables may be used in order to predict the criterion variable more accurately.

14. The regression equation for multiple regression is an extension of the equation for simple regression. It includes an intercept a (the constant), and a slope b (the regression coefficient) for each of the predictor variables. With two predictors (X_1 and X_2) the multiple regression prediction equation is:

$$Y' = b_1 X_1 + b_2 X_2 + a$$

15. The combined correlation of the predictor variables with the criterion variable is called **multiple R**.

16. R^2 (which is similar to r^2 in simple regression) indicates the proportion of the variation in Y (the criterion variable) that can be accounted for by the variation of the combined predictor variables.

Part Five

Inferential Statistics

Chapter 10

t Test

Many researchers are concerned with comparing groups to each other. Their studies are conducted to determine whether the differences between groups are statistically significant, or whether they could have occurred simply by chance. When two means are being compared with each other, the statistical test that is used is a *t* **test**. For example, research studies may compare an experimental group to a control group, men to women, or pretest to posttest scores. The numbers used for the comparison are the means of the two groups or set of scores. The scores used to compute the means should be measured on an interval or ratio scale and be derived from the same measure (e.g., the same test or an alternate form of the same test). As we analyze research data, we should keep in mind that small differences are expected even among members of the same group. These differences may occur due to sampling error and are considered chance differences.

You might ask how we can distinguish between differences due to sampling error and "real" differences. At what point do we say that the difference is too large to be attributed to sampling error, and that it probably indicates a real difference? Unfortunately, there are no standards or cutoff scores. After obtaining the means, we cannot simply "eyeball" them and determine whether they are similar or different. A difference of 2 points between means may be defined as statistically significant in some cases, but not in others. The group means and variances, in addition to the sample size, all play a role in determining whether the difference between the means is a "real" difference.

The *t* test is based on the *t distribution* that was developed in 1908 by W. S. Gosset, who worked for a brewery in Dublin. Since employees were not allowed to publish in journals, Gosset used the pseudonym "Student" in an article he sent to a journal, and the *t* distribution became known as "Student's *t* distribution."

Hypotheses for *t* Tests

Predictions of outcomes in studies that are using the *t* test state the expected differences between two means. These predictions are the research hypotheses and they reflect what the researcher hypothesizes about the nature of the differences between the means.[1] The alternative hypothesis (i.e., the research hypothesis), represented by the symbol H_A or H_1 predicts whether there would be a statistically significant difference between the two means being compared. For example, the alternative hypothesis may be a *directional hypothesis* that predicts that the baggage handlers at an airline using a new computer chip to track luggage (experimental group) would score higher on a consumer satisfaction survey, compared to baggage handlers at another airport not using the new tracking chip (control group). The alternative directional hypothesis is:

---

1. **A HINT:** See Chapter 2 for a full explanation of alternative (research) and null hypotheses.

$$H_A : \text{Mean}_E > \text{Mean}_C$$

Where H_A =   Research hypothesis (the alternative hypothesis)

           Mean_E =   Mean of the experimental group

           Mean_C =   Mean of the control group

A *nondirectional hypothesis* in a *t* test predicts that there would be a difference between the two means, but the direction of the outcome is not specified. The nondirectional hypothesis is:

$$H_A : \text{Mean}_1 \neq \text{Mean}_2$$

For example, we may predict differences in attitudes toward technology between men and women, but due to inconclusive results in previous studies, we are unable to predict which of the two groups will have a more positive attitude.

Occasionally, the research hypothesis is not stated as directional or nondirectional, but in a null form. That is, we predict that there will be no difference between the means. This is not very common in business research, but in cases where the research hypothesis is stated as null, that hypothesis is considered nondirectional.

The *null hypothesis* (H_0) in the *t* test states that any observed difference between the means is too small to indicate a real difference between them and that such difference is probably due to sampling error. In other words, the null hypothesis always predicts no difference between the means.[2] In symbols, the null hypothesis is:

$$H_O : \text{Mean}_1 = \text{Mean}_2$$

Or: $H_0 : \text{Mean}_1 - \text{Mean}_2 = 0$

The null hypothesis is submitted to a statistical test. Based on the results of this statistical test, we decide whether to retain or reject the null hypothesis. Since the null hypothesis always predicts no difference, there is no need to formally state it when the research hypotheses are presented. However, when the research (alternative) hypothesis predicts no difference between the means, then it should be stated and included among the other study's research hypotheses.

In order to calculate the *t* test value, a score should be obtained for each subject or case. The scores are then used to calculate the *t* value. After calculating a *t* value, the next step is to consult a table of critical values for the *t* distribution to determine the level of significance (*p* value) of the obtained *t* value. To use the table, the researcher needs to know whether to use the critical values listed under the **one-tailed** or **two-tailed test**. Without going into a lengthy explanation about what the tails are, remember the following rule: If your research hypothesis is *directional* and you predict which mean will be higher, use the *one-tailed test*. If your research hypothesis is *nondirectional* and you predict a difference between the means but do not specify which mean will be higher, use a two-tailed test. If your hypothesis is stated as *null* and you predict no difference between the means beyond what might be expected purely by chance, use the *two-tailed test*. When in doubt, use the *two-tailed test* that is considered more conservative.[3]

---

2 . **A HINT**: This is true when we use the *t* test. You may remember that in correlational studies, the null hypothesis always predicts no relationship between the variables being correlated (i.e., it predicts a correlation that does not differ significantly from zero; see Chapter 8).

3. **A HINT**: A conservative test or a conservative decision generally reduces the chance of making a *Type I error*.

Today, with computer statistical software packages readily available, there is no need to compute the *p* value by hand. The computer program will provide the appropriate *t* value and *p* value, along with the two means and standard deviations. However, you will still need to know whether to use a one-tailed or two-tailed test.

Using the *t* Test

A *t* test is used to compare two means in three different situations:

1. **Independent samples *t* test**. The two groups whose means are being compared are independent of each other. A typical example is a comparison of experimental and control groups.

2. **Paired samples *t* test** (also called a *t* test for **dependent, matched,** or **correlated** samples). The two means represent two sets of scores that are paired. A typical example is a comparison of pretest and posttest scores obtained from one group of people.

3. **A single sample *t* test**. This *t* test is used when the mean of a sample is compared to the mean of a population. For example, we may use the Graduate Management Admission Test (GMAT) scores of accounting graduate students (the sample) to test whether they are significantly different from the overall mean GMAT in the whole business school (the population).

In using the *t* test, it is assumed that the scores of those in the groups being studied are normally distributed and that the groups were randomly selected from their respective populations. In studies conducted in business and behavioral sciences, it is quite difficult to satisfy these requirements. However, empirical studies have demonstrated that we can use the *t* test even if the assumptions are not fully met.

t Test for Independent Samples

The *t* test for independent samples is used extensively in experimental designs and in causal-comparative (ex-post facto) designs when means from two groups are being compared. There are several assumptions underlying this test:

1. The groups are independent of each other.
2. A subject may appear in only one group.
3. The two groups come from two populations whose variances are approximately the same. This assumption is called the **assumption of the homogeneity of variances**. We compare the two variances to determine if there is a statistically significant difference between them. When the two groups are approximately the same size, there is no need to test for the homogeneity of variances.

To test for the assumption of the homogeneity of the variances, we divide the larger variance by the smaller variance and obtain a ratio, called the **F value** (or **F ratio**). When the *F* value is statistically significant we cannot assume that the variances are equal. In order to decide whether the *F* value is statistically significant, we consult the *table of critical values* of the *F* distribution.[4] A *test for the equality of variances*, such as the *Levene's test*, is used to test the significance of the *F* value. An *F* value that is *not* statistically significant ($p > .05$) indicates that the assumption for the homogeneity of variances is *not* violated and, therefore, *equal variances* can be assumed. On the other hand, a *significant F* value ($p < .05$) indicates that the assumption for the homogeneity of variances was violated. The *t* test statistical results are adjusted for the *unequal variances*.

---

4. **A HINT:** The table of critical values of the *F* distribution is also used in the analysis of variance (ANOVA) that is discussed in Chapter 11.

It is unlikely that you will have to do the computations by hand in order to test for the homogeneity of variances. Most statistical software packages (such as SPSS) include the level of significance of the F value in their report of results for independent samples t test.

The t test is considered a **robust** statistic. Therefore, even if the assumption of the homogeneity of variance is not fully met, the researcher can probably still use the test to analyze the data. As a general rule, *it is desirable to have similar group sizes, especially when the groups are small*.[5]

An Example of a t Test for Independent Samples

An advertising agency wants to convince one of its consumer goods clients that implementing a new ad campaign would significantly improve sales of its shampoo product. The consumer goods company selects 20 cities at random and assigns 10 to an experimental group and 10 to a control group.[6] Consumers in the experimental group of cities are exposed to the new advertising campaign and the consumers in the other 10 cities continue to be exposed to the old ad campaign. After one month, the company records the percentage increases in sales in both groups of cities. The consumer goods company then conducts a t test for independent samples to compare the percentage increases in sales of the cities in Group 1 (Experimental Group E) to those in Group 2 (Control Group C).

The study's research (alternative) hypothesis (H_A) is directional and can be described as:

$$H_0 : \mu_E = \mu_C$$

Note that μ, the symbol for the population mean, is used in writing hypotheses. Remember that although we may conduct our studies using *samples*, we are testing hypotheses about *populations*.

The null hypothesis states that there is no significant difference between the two means. A study is then designed to test the null hypothesis and to decide whether it is tenable. The null hypothesis is:

$$H_0 : \mu_E = \mu_C$$

The t value is computed using this formula:

$$t = \frac{\overline{X}_1 - \overline{X}_2}{\sqrt{\frac{(n_1 - 1)S_1^2 + (n_2 - 1)S_2^2}{n_1 + n_2 - 2} \left(\frac{1}{n_1} + \frac{1}{n_2} \right)}}$$

Where			
$\overline{X}_1$	=	Mean of Group 1 (Experimental)
$\overline{X}_2$	=	Mean of Group 2 (Control)
S_1^2	=	Variance of Group 1
S_2^2	=	Variance of Group 2
n_1	=	Number of cities in Group 1 (experimental group)
n_2	=	Number of cities in Group 2 (control group)

5. **A HINT:** As in several other statistical tests, researchers usually try to have a group of at least 30. Larger samples are more stable and require a smaller t value (compared with smaller samples) to reject the null hypothesis.
6. **A HINT:** Although in real-life studies researchers try to have larger sample sizes, in this chapter (as well as in other chapters) we are using small sample sizes to simplify the computations in the examples given.

As we can see, in addition to the difference between the two means (the numerator), the formula also includes the two sample sizes and the two variances (the denominator). This can explain why we cannot simply look at the difference between the two means and decide whether that difference is statistically significant. The number of scores in each group and the variability of these scores also play a role in the *t* test calculations. The difference between the means is viewed in relation to the spread of the scores. When the spread is small (a low variance), even a small difference between the means may lead to results that are considered statistically significant. With a larger spread (a higher variance), a relatively large difference between the means may be required in order to obtain results that are considered statistically significant.

After finding the *t* value, it is then compared to appropriate critical values in the **t test table of critical values**. When the obtained *t* test *exceeds* its appropriate critical value, the null hypothesis is *rejected*. This allows us to conclude that there is a high level of probability that the difference between the means is notably greater than zero and that a difference of this magnitude is unlikely to have occurred by chance alone.

The following are percentage increases in sales and the computations of the *t* value for the cities in the experimental and control groups in the study that was conducted by the consumer goods company (Table 10.1):

Table 10.1   **Percentage Sales Increases of Experimental Group**
($n = 10$) and Control Group ($n = 10$)

Experimental Group 1- % Increase	Control Group 2 - % Increase
26	19
27	24
21	18
31	23
21	22
25	24
29	32
32	29
34	15
23	20

Mean	$\overline{X}_1 = 26.90$	$\overline{X}_2 = 22.60$
SD	$S = 4.56$	$S = 5.08$
Variance	$S_1^2 = 20.77$	$S_2^2 = 25.83$
Sample size	$n_1 = 10$	$n_2 = 10$

$$t = \frac{26.90 - 22.60}{\sqrt{\frac{(10-1)(20.77) + (10-1)(25.83)}{10 + 10 - 2}\left(\frac{1}{10} + \frac{1}{10}\right)}} =$$

$$\frac{4.30}{\sqrt{\frac{(186.93 + 232.47)}{18}(0.10 + 0.10)}} = \frac{4.30}{\sqrt{\frac{419.40}{18}(0.20)}} =$$

$$= \frac{4.30}{\sqrt{(23.30)(0.20)}} = \frac{4.30}{\sqrt{4.66}} = \frac{4.30}{2.159} = 1.992$$

When the two groups have the same number of observations (i.e., $n_1 = n_2$), a simpler formula can be used to compute the t value:

$$t = \frac{\overline{X}_1 - \overline{X}_2}{\sqrt{\dfrac{S_1^2}{n_1} + \dfrac{S_2^2}{n_2}}} = \frac{26.90 - 22.60}{\sqrt{\dfrac{20.77}{10} + \dfrac{25.83}{10}}} = \frac{4.30}{\sqrt{2.077 + 2.583}} =$$

$$= \frac{4.30}{\sqrt{4.66}} = \frac{4.30}{2.159} = 1.992$$

Next, we consult the table of critical values (below). To do so, we need to determine the appropriate degrees of freedom (df). As was described before (see Chapter 2), df are typically n-1 (number of cases minus 1). In a t test for independent samples with two groups, df are computed as:

$$(n_1 - 1) + (n_2 - 1) \qquad \text{or} \qquad n_1 + n_2 - 2$$

In our example, we have 10 cities in the experimental group and 10 cities in the control group. Therefore, our degrees of freedom are 18. Following is a section from the table of the critical values of the t distribution for 18 degrees of freedom (df=18).

	Table 10.2		Partial Distribution of t Table for 18 df			
	Level of significance for one-tailed test					
p values	.10	.05	.025	.01	.005	.0005
	Level of significance for two-tailed test					
p values	.20	.10	.05	.02	.01	.001
Critical Values	1.330	1.734	2.101	2.552	2.878	3.922

The hypothesis for this study was directional; therefore, we use the *one-tailed test*. Using the table, we locate the appropriate critical value under a p value of .05 for 18 degrees of freedom. We determine that the critical value is 1.734. (This is listed as $t_{crit(.05,18)} = 1.734$, with .05 showing the p level, and 18 indicating the df.) Our obtained t value of 1.992 *exceeds* the corresponding critical value of 1.734; therefore, we move to the right to the next column and check the critical value at p=.025, which is 2.101 ($t_{crit(.025,18)} = 2.101$). Our obtained value of 1.992 *does not* exceed this critical value. Therefore, we report our results to be significant at the p<.05 level, which is the last critical value that we did exceed. This information can be summarized as:

Obtained t = 1.992	df = 18	Use a one-tailed test
$t_{crit(.05,18)}$ = 1.734	$t_{crit(.025,18)}$ = 2.102	Reject the null hypothesis at p <.05

Our decision is to reject the null hypothesis. Such a large difference between the two groups could have occurred by chance alone less than 5% of the time.[7] The hypothesis stated by the consumer goods company is confirmed: the

---

7.   **A HINT**: You should remember, though, that we are making our statistical decision in terms of *probability*, not *certainty*. Rejecting the null hypothesis at p <.05 means that there is a possible error associated with this decision.

cities where the new advertising campaign was implemented had significantly higher increases in sales than the cities in the control group. Based on the result of this study, we can be at least 95% confident that the increase in sales was not attributed strictly to chance.

As was discussed in Chapter 2, reporting the *statistical* significance of an experimental study should be followed by an evaluation of its *practical* significance. In addition to inspecting and evaluating the difference between the means of the two groups, we can use the index of effect size (ES) to evaluate the practical significance of our study and the effectiveness of the intervention (the advertising campaign, in our example). As you recall, the effect size is calculated using this formula:

$$ES = \frac{Mean_{EXP} - Mean_{CONT}}{SD_{CONT}}$$

The numerator is the difference between the means of the experimental and control groups and the denominator is the standard deviation of the control group. Using the data in Table 10.1, we can now calculate the effect size as follows:

$$ES = \frac{26.90 - 22.60}{5.082} = \frac{4.30}{5.082} = .85$$

Our ES is .85, which is considered high. This ES confirms that the difference of 4.30 points between the two groups is practically significant, in addition to being statistically significant (at $p<.05$). This information seems to support the statistical significance that indicated that the likelihood of getting such a large difference between the two groups purely by chance is less than 5%. We can conclude that implementing the new advertising campaign was effective and can bring about a statistically and practically significant increase in sales. Of course, this was a study with a very small sample size, and the consumer goods company may need to repeat the study with larger samples to really convince the accountants and the CEO!

t Test for Paired Samples

A *t* test for paired samples is used when the two means being compared come from two sets of scores that are related to each other. It is used, for example, in experimental research to measure the effect of an intervention by comparing the posttest to the pretest scores. The most important requirement for conducting this *t* test is that the two sets of scores are *paired*. In studies using a pretest-posttest design, it is easy to see how the scores are paired: they belong to the same individuals. It is assumed that the two sets of scores are normally distributed and that the samples were randomly selected.

For example, we may want to investigate if exposure to advertising for a specific brand name athletic shoe increases the likelihood that teenagers will purchase that brand of shoes. In such a study, we can administer a purchase intent survey to teenagers before seeing advertising for the shoes and use the scores on the survey as our pretest. The second set of scores would be the teenagers' purchase intent after watching the ads for the brand name shoes.

To compute the paired samples *t* test, we need to first find, for each person, the difference (*D*) between the two scores (e.g., between pretest and posttest) and sum up those differences (ΣD). Usually, the lower score (e.g., pretest) is subtracted from the higher score (e.g., posttest) so D values are positive. We also need to compute the sum of the squares of the differences (ΣD^2). The t value is computed using this formula:

$$t = \frac{\Sigma D}{\sqrt{\dfrac{n(\Sigma D^2) - (\Sigma D)^2}{n-1}}}$$

Where	ΣD	=	Sum of the difference scores (D)
	ΣD^2	=	Sum of the squared differences (D^2)
	n	=	Number of pairs of scores

The example that follows demonstrates the computation of a paired samples t test. To simplify the computations, we use 8 sets of scores only. In conducting real-life research, it is recommended that larger samples (30 or more) be used.

An Example of a t Test for Paired Samples

A CPA preparation firm needs to convince prospective students to take its CPA prep course. It wants to show prospective students that taking the test will significantly improve their chances of passing the CPA exam. The class lasts six weeks and involves various activities in the classroom and homework. The instrument used to assess the effectiveness of the class is comprised of questions from past CPA exams, and scores can range from 0 to 40. All of those who complete the CPA prep course are tested before the start of the program and then again at the end of the course. A t test for paired samples is used to test the hypothesis that students' scores on the 40-point test would improve significantly on the posttest, compared with their pretest scores. The hypothesis is:

$$H_A: \text{Mean}_{POST} > \text{Mean}_{PRE}$$

Table 10.3 shows the numerical data we need to compute the t value for the 8 students selected at random from the class. The table shows the pretest and posttest scores for each student, as well as the means on the pretest and posttest. The third column in the table shows the difference between each pair of scores (D) and is created by subtracting the pretest from the posttest for each participant. The gain scores are then squared and recorded in the fourth column (D^2). The scores in these last two columns are added up to create ΣD and ΣD^2, respectively. These are used in the computation of the t value which follows Table 10.3.

Table 10.3   Pretest and Posttest Scores of Eight Students

Pretest X_1	Posttest X_2	Posttest – Pretest D	(Posttest–Pretest)^2 D^2
30	31	+1	1
31	32	+1	1
34	35	+1	1
32	40	+8	64
32	32	0	0
30	31	+1	1
33	35	+2	4
34	37	+3	9
$\bar{X}_1 = 32.00$	$\bar{X}_2 = 34.13$	$\Sigma D = 17$	$\Sigma D^2 = 81$

The following computations show that the obtained *t* value is 2.37. This obtained *t* value is then compared to the values in the abbreviated table of critical values of the *t* distribution (Table 10.4).

$$t = \frac{\sum D}{\sqrt{\dfrac{n\sum D^2 - (\sum D)^2}{n-1}}} = \frac{17}{\sqrt{\dfrac{(8)(81)-(17)^2}{8-1}}} =$$

$$= \frac{17}{\sqrt{\dfrac{359}{7}}} = \frac{17}{\sqrt{51.29}} = \frac{17}{7.16} = 2.37$$

Obtained *t* = 2.37   *df* = 7   Use a one-tailed test

$t_{crit(.05,7)} = 1.895$ $t_{crit(.025,7)} = 2.365$ $t_{crit(.01,7)} = 2.998$

Reject the null hypothesis at *p* < .025

	Table 10.4		Partial Distribution of *t* Table for 7 *df*			
	Level of significance for one-tailed test					
p values	.10	.05	.025	.01	.005	.0005
	Level of significance for two-tailed test					
p values	.20	.10	.05	.02	.01	.001
Critical Values	1.415	1.895	2.365	2.998	3.499	5.405

In this study, our hypothesis was directional and we predicted that the posttest scores would be significantly higher than the pretest scores. Therefore, we need to use the critical values under the one-tailed test. Our obtained *t* value of 2.37 *exceeds* the critical values under *p*=.05 (one-tailed) and the critical value of *p*=.025, which is 2.365 (although the obtained value is almost the same as the critical value under *p*=.025). However, the obtained *t* value *does not exceed* the critical value under p<.01, which is 2.998. Therefore, we reject the null hypothesis that states that there is no difference between the pretest and the posttest scores and report our results as significant at the *p*<.025 level. The chance that these results were obtained purely by chance is less than 2.5%. We confirm the research hypothesis that predicted that the posttest mean score would be significantly higher than the pretest mean score. According to this study, the prep course was effective in increasing the chances of those taking the course to score higher on the CPA exam.

The data in Table 10.3 show that there was an increase of a little over 2 points (2.13 points, to be exact) in the students' mean score from pretest to posttest. We confirmed that the results were statistically significant and want now to evaluate their practical significance by calculating the effect size (see Chapter 2). The appropriate formula for calculating ES for gain scores is:

$$ES = \frac{Mean_{POSTTEST} - Mean_{PRETEST}}{SD_{GAIN}}$$

The numerator is the difference between the pretest and posttest means. The denominator is the standard deviation of the gain score (SD_{GAIN}). A number of computer software programs can provide us with the standard deviation for the gain scores, or we can also compute this SD by hand using the gain scores found in Table 10.3. Entering the means for the pretest, posttest, and SD_{GAIN} into the equation, we get an effect size (ES) of .84.

$$ES = \frac{34.13 - 32.00}{2.53} = \frac{2.13}{2.53} = .84$$

An effect size of .84 is considered high in business. Our data seem to indicate that the class to increase the skills needed to pass the CPA exam was effective. However, those conducting the research or those considering taking the course will still have to determine for themselves whether the course is worthwhile. The question to be asked is whether an increase of 2.13 points (out of 40 possible points on the scale) is worth the investment of time, money, and effort.

t Test for a Single Sample

Occasionally, a researcher is interested in comparing a single group (a sample) to a larger group (a population). For example, a Chief Financial Officer (CFO) in a company may want to confirm that the mean salary her company is paying its accountants is competitive with accountants in the rest of the U.S. In order to carry out this kind of a study the researcher *must* know prior to the start of the study the mean value of the population. In this example, the mean score of the population is the overall mean of the salaries of *all* accountants in the U.S.

An Example of a Single Sample *t* Test

A plant manager is concerned that his factory's quality scores are not staying ahead of the industry norm. To test whether his quality is the same as the industry average, he conducted a *t* test for a single sample. The quality scores used in this industry are calculated using a formula that includes the number of good parts manufactured, the number of defects, and level of waste. The scores on an industry-wide quality scale can range from zero to 150. In this example, we consider the industry average to be the population to which we compare the mean of the factory. The mean quality score of the population is 110 (μ=110) and this factory's mean quality score is 112.20 (= 112.20). The hypothesis states that there is no difference in the mean of the plant's quality score (the sample) and the mean quality score for the industry (the population).

$$H_A : Mean_{FACTORY} = Mean_{INDUSTRY}$$

The formula for *t* test of a single sample is:

$$t = \frac{\bar{X} - \mu}{S_{\bar{x}}}$$

Where X =   Sample mean
            μ =   Population mean
            $S_{\bar{x}}$ =   Standard error of the mean

To find $S_{\bar{X}}$ we use this formula:

$$S_{\bar{X}} = \frac{S}{\sqrt{n}}$$

Where $S_{\bar{X}}$ = Standard error of the mean

S = Sample standard deviation

n = Number of individuals in the sample

In order to test his hypothesis, the manager randomly selects quality scores for 10 weeks of this year's production runs. These quality scores are listed in Table 10.5, followed by the computation of the *t* value.

Table 10.5   **Quality Scores of 10 Production Runs**

Scores

115	118
135	113
105	98
107	120
112	99

ΣX	=	1122
$\bar{X}$	=	112.20
S (SD)	=	10.94

$$S_{\bar{x}} = \frac{S}{\sqrt{n}} = \frac{10.94}{\sqrt{10}} = \frac{10.94}{3.16} = 3.46$$

$$t = \frac{\bar{X} - \mu}{S_{\bar{x}}} = \frac{112.20 - 110}{3.46} = \frac{2.20}{3.46} = 0.64$$

Obtained $t = 0.64$ $df = 9$ Use a two-tailed test

$t_{\text{crit}(.05,9)} = 2.262$ Retain the null hypothesis

The obtained *t* value of 0.64 is then compared to the critical values for 9 degrees of freedom using the following abbreviated table of critical values for the *t* distribution in Table 10.6. Our research hypothesis predicted that there will be no difference between the sample and population means; therefore, we use the two-tailed test.

Table 10.6		Abbreviated Distribution of *t* table for 9 *df*			
Level of significance for one-tailed test					
p values .10	.05	.025	.01	.005	.0005
Level of significance for two-tailed test					
p values .20	.10	.05	.02	.01	
1.383	1.833	2.262	2.821	3.250	4.781

The obtained *t* value of 0.64 *does not exceed* the critical value under $p=.05$, which is 2.262 (see Table 10.6 above). Thus, the null hypothesis is retained.[8] Based on these results, the manager concludes that there is no statistically significant difference between his quality scores and the quality scores for the industry. In fact, his mean quality score (mean = 112.20) is actually slightly higher than the mean score of the industry (mean = 110). The research hypothesis that was stated in a null form (i.e., predicting no difference between the two means) is confirmed.

Summary

1. The *t* test is used to compare two means to determine whether the difference between them is statistically significant.

2. The *t* test requires data measured on an *interval* or a *ratio* scale.

3. A *directional* hypothesis in *t* test predicts which of the two means is going to be higher (H_A: Mean$_1$ > Mean$_2$).

4. A *nondirectional* hypothesis in *t* test that predicts a difference between the two means, but does not specify which mean will be higher (H_A: Mean$_1$ ≠ Mean$_2$).

5. The *null* hypothesis in *t* test states that any observed difference between the means is too small to indicate a real difference between them and that such difference is probably due to sampling error. In other words, the null hypothesis always predicts no difference between the means beyond what might happen purely by chance (Ho: Mean$_1$ = Mean$_2$).

6. When using the table of critical values, *directional* hypotheses are tested using the **one-tailed** test, and *nondirectional* hypotheses are tested using the **two-tailed** test. When in doubt, use the two-tailed test, which is more conservative.

7. The *t* test can be used to compare means from: (a) two independent groups, (b) two paired groups, and (c) a

---

8. **A HINT:** There is a certain level of error in any statistical decision. When we decide to retain the null hypothesis, we risk making a *Type II* error (see Chapter 2), where we retain the null hypothesis when, in fact, it should be rejected. In our particular example, however, *t* value is very small ($t = 0.53$) and our decision to retain the null hypothesis is likely to be the proper decision.

single sample and a population.

8. In using the *t* test, it is assumed that the scores of those in the groups being studied are normally distributed and that the groups were randomly selected from their respective populations. In studies conducted in business, it is sometimes difficult to satisfy these requirements. However, empirical studies have demonstrated that we can use the *t* test even if these assumptions are not fully met.

9. The *t* **test for independent samples** is used when the two groups whose means are being compared are independent of each other. For example, this *t* test may be used to compare the means of experimental and control groups.

10. When conducting the *t* test for independent samples, it is assumed that the two groups being compared come from two populations whose variances are approximately the same. This assumption is called the assumption of the **homogeneity of variances**. We can compare the two variances to check if they are not significantly different from each other.

11. The *t* test is considered a **robust** statistic; therefore, even if the assumption about the homogeneity of the variance is not met, the researcher can still safely use the test to analyze the data. As a general rule, it is desirable to have similar group sizes, especially when the groups are small.

12. The *t* **test for paired sample** (also called a *t* test for **dependent, matched,** or **correlated** samples) is used when the means come from two sets of paired scores.

13. The *t* **test for a single sample** is used when the mean of a sample is compared to the mean of a population. In order to carry out this kind of a study the researcher *must* know prior to the start of the study the mean value of the population.

14. In studies that compare two means, an *effect size (ES)* may also be calculated, in addition to the *t* value. The ES is a ratio that can be calculated to assess whether a difference between two means is *practically significant*, or *practically important*. In calculating an effect size, the numerator is the difference between the two means is the denominator is a standard deviation.

Chapter 11

Analysis of Variance

The **analysis of variance (ANOVA)** test is used to compare the means of two or more independent samples and to test whether the differences between the means are statistically significant. ANOVA, which was developed by R. A. Fisher in the early 1920s, can be thought of as an extension of the t test for independent samples. However, a t test can compare only two means, whereas an ANOVA can compare two or more means simultaneously.[1]

Suppose, for example, that we want to compare five groups. If a t test is used, we have to repeat it 10 times to compare all the means to each other. We have to compare the mean from group 1 with the means from groups 2, 3, 4, and 5; and the mean from group 2 with the means from groups 3, 4, and 5; and so on. Every time we do a t test, there is a certain level of error that is associated with our decision to reject the null hypothesis and the error is compounded as we repeat the test over and over. The main risk is that we may make a Type I error; that is, reject the null hypothesis when in fact it is true and should be retained (see Chapter 2). By comparison, when we use ANOVA to compare the five means simultaneously, the error level can be kept at the .05 level (5%). In addition to keeping the error level at a minimum, performing one ANOVA procedure is more efficient than doing a series of t tests.

In ANOVA, the *independent variable* is the categorical variable that defines the groups that are compared (e.g., gender, age cohort, or marital status). The *dependent variable* is the measured variable whose means are being compared (e.g., sales, level of job satisfaction, or level of compliance with accounting standards).

There are several assumptions for ANOVA: (a) the groups are independent of each other, (b) the dependent variable is measured on an interval or ratio scale, (c) the dependent variable is normally distributed in the population, (d) the scores are random samples from their respective populations, and (e) the variances of the populations from which the samples were drawn are equal (the assumption of the *homogeneity of variances*). The first two assumptions (*a* and *b*) must always be satisfied. Assumptions *c* and *d* are often difficult to satisfy in business and behavioral sciences. However, even if we cannot determine that random sampling was used, we can generally satisfy the requirement that the samples are not biased. ANOVA is considered a *robust* statistic that can stand some violation of the third and fourth assumptions, and empirical studies show that there are no serious negative consequences if these assumptions are not fully met. The last assumption (*e*) can be tested using special tests, such as the F test, in which the largest variance is divided by the smallest variance.

The test statistic in ANOVA is called the F statistic. The **F value** (or **F ratio**) is computed by dividing two variance estimates by each other.[2] If the F ratio is statistically significant (i.e., if $p<.05$) and if there are three or more groups in the study, then a pair-wise comparison is done to assess which two means are significantly different from each other.[3]

---

1. **A HINT**: Although ANOVA can be used with two or more groups, most researchers use the independent samples t test when the study involves two independent groups and ANOVA is used when there are three or more independent groups in the study.
2. **A HINT**: Even though the ANOVA test is designed to compare means, the samples' *variances* and *variance estimates* are used in the computations of the F ratio.
3. **A HINT**: An example of the pair-wise comparison is discussed later in this chapter in the section titled *Post Hoc Comparison*.

When one independent variable is used, the test is called a **one-way analysis of variance (one-way ANOVA)**. To illustrate, let's look at a study to test the effect of three levels of credentials on the performance evaluation scores of first-year accountants at a major corporation. In this study, the level of credentials is the *independent* variable and the first-year accountants' performance evaluation scores are the *dependent* variable. Three levels of credentials are tested: bachelor's degree only, bachelor's degree having passed the CPA exam, and MBA. At the end of the year, the performance evaluation scores of the accountants who worked for the firm with these three different levels of credentials are compared to each other using a one-way ANOVA. The *F* ratio would be used to assess whether there are significant differences in the mean scores of the three groups of first-year accountants' performance.

The ANOVA test can be applied to studies with more than one independent variable. For example, we can study the relationship between *two* independent variables and a dependent variable. When there are two independent variables, the design is called a **two-way analysis of variance**. In general, when two or more independent variables are studied in ANOVA, the design is called a **factorial analysis of variance**.

Let's go back to our example of the three levels of credentials and first-year accountants' performance evaluation scores. The three levels of credentials were bachelor's degree only, bachelor's degree having passed the CPA exam, and MBA. Based on prior research and our own experience, suppose we believe that the first-year accountants' gender also makes a difference and that performance evaluation scores of first-year male and female accountants would differ depending on the level of credentials. A two-way ANOVA (i.e., a factorial ANOVA) can be used to explore the effect of the two independent variables (level of credentials and gender) on the dependent variable (performance evaluation scores). In addition, using the two-way ANOVA would allow us to study the *interaction* between all variables. In the factorial ANOVA test, the **interaction** refers to a situation where one or more levels of the independent variable have a different effect on the dependent variable when combined with another independent variable. For example, we may find that men with MBAs have a higher performance evaluation score compared with men in the other two groups whereas women have higher performance scores when they have a bachelor's degree and CPA compared with women in the other two groups.

An independent variable must have at least two **levels** (or conditions). In our example, the variable of gender has two levels and the variable of credentials has three levels. To further explain the concept of levels of independent variables, let's look at the following examples of two-way and three-way ANOVA tests.

Suppose we conduct a study of finance managers to investigate the relationship between two independent variables, gender and industries in which they work and their effect on the managers' likelihood of increasing their companies' debt structure (the dependent variable). The variable of gender has 2 levels (female and male) and the variable of industry in which they work has 3 levels (manufacturing, service, non-profit). The design of the study is indicated as a *2x3 factorial ANOVA* (or a *two-way ANOVA*). Assume we want to add a third independent variable to our study, such as the size of the company at which each manager works. We would assign a code of 1-3 to each manager depending on her or his company's size. A code of 1 would be assigned to managers whose companies have more than 5,000 employees; a code of 2 would be assigned to managers whose companies have between 500 and 5,000 employees; a code of 3 would be assigned to managers whose companies' have fewer than 500 employees. The design of our study would be: 2x3x3 ANOVA.

One-Way ANOVA

Conceptualizing the One-Way ANOVA

ANOVA studies three types of variability that are called the **sum of squares** (abbreviated as SS). They are:

1. **Within-groups sum of squares (SS_W)**, which is the variability within the groups.

2. **Between-groups sum of squares groups (SS_B)**, which is the average variability of the means of the groups around the total mean. (The total mean is the mean of all the scores, combined.) It may also be called **among-groups sum of squares**, abbreviated as SS_A.

3. **Total sum of squares (SS_T)**, the variability of all the scores around the total mean.[4]

The *total* sum of squares is equal to the combined *within* groups sum of squares and the *between* groups sum of squares:

$$SS_T = SS_W + SS_B$$

Figure 11.1 illustrates the different sum of squares and shows that sum of square within (SS_W) plus sum of square between (SS_B) equal the total sum of square (SS_T). In this figure, X_1 is the score of an individual in Group 1; $\overline{X}_1$ is the mean of Group 1; and $\overline{X}_T$ is the total mean.

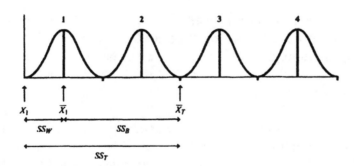

Figure 11.1   A graph showing the three sums of squares: SS_W, SS_B, and SS_T

After finding the sums of squares, the next step is to compute the variance estimates. The variance estimates are called the **mean squares**, abbreviated as **MS**. The mean squares are found by dividing the sum of squares (SS) by the appropriate degrees of freedom. This process is similar to the computation of a variance for a single sample where we divide the sum of the squared distances of scores around their means (the sum of squared deviations) by n-1 (see Chapter 5). In ANOVA, the degrees of freedom are used as the denominator in place of N-1 when computing the variance.

---

4. **A HINT**: Think of it as combining the scores from all the groups to create one large group, and computing the variability of this group around the total mean.

The degrees of freedom for *within* (df_W) are found by subtracting the number of groups in the study (i.e., K) from the total number of individuals in the study $(N\text{-}K)$. The degrees of freedom for *between* (df_B) are found be subtracting 1 from the total number of groups $(K\text{-}1)$. The degrees of freedom associated with the *total* variance estimate (df_T) are the total number of scores minus 1 $(N\text{-}1)$. The df_T is equal to the combined degrees of freedom for the within mean square (df_W) and between mean squares (df_B). For example, let's say we have three groups in our study with 5 people in each group. The total number of people in the study is 15 (3x5=15). In this example, $K=3$, $n=5$, and $N=15$. We calculate the degrees of freedom as follows:

$$df_W = N - K = 15 - 3 = 12$$
$$df_B = K - 1 = 3 - 1 = 2$$
$$df_T = N - 1 = 15 - 1 = 14$$

To compute the F ratio, we need only two variance estimates, MS_W and MS_B. Therefore, there is no need to compute the total mean square (MS_T). The formulas for computing MS_W and MS_B are:

$$MS_W = \frac{SS_W}{N - K} \qquad MS_B = \frac{SS_B}{K - 1}$$

The MS_W (also called the **error term**) can be thought of as the average variance to be expected in any normally distributed group. The MS_W serves as the denominator in the computation of the F ratio. To compute the F ratio, the *between*-group mean square (MS_B) is divided by the within-group mean square (MS_W).

$$F = \frac{MS_B}{MS_W}$$

MS_B, the numerator, increases as the differences between the group means increase; therefore, greater differences between the means result in a higher F ratio. Additionally, since the denominator is the within-group mean square, when the groups are more homogeneous and have lower variances, the MS_W tends to be smaller and the F ratio is likely to be higher. Two figures are presented to illustrate the role of group means (Figure 11.2) and variances (Figure 11.3) in the computations of the F ratio in ANOVA.

Part a and *Part b* in Figure 11.2 show the distributions of scores of several groups. The variances of the three groups in *Part a* (Groups 1, 2, and 3) are about the same as the variances of the three groups in *Part b* (Groups 4, 5, and 6). However, the means of the three groups in *Part a* are farther apart from each other, compared with the means of the three groups in *Part b*. If asked to predict which part of Figure 11.2 would yield a higher F ratio, we would probably choose *Part a*, where the three groups do not overlap and the means are quite different from each other. By contrast, the means of Groups 4, 5, and 6 are closer to each other, and the three distributions overlap.

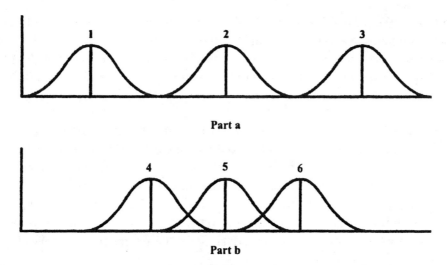

Figure 11.2   A graph showing three groups with different means and similar variances (*Part a*),
and a graph showing three groups with more similar means and similar variances (*Part b*)

Next, let's turn our attention to Figure 11.3 where two sets of distributions are presented in *Part a* (Groups A, B, and C) and in *Part b* (Groups D, E, and F). Notice that the three groups in *Part a* have the same *means* as the three groups in *Part b*, but the *variances* in both parts are different. The variances of the three groups depicted in *Part a* are very low; that is, the groups are homogeneous with regard to the characteristic being measured. By comparison, the variances of the three groups depicted in *Part b* are higher, with a wider spread of scores in each group. If we were to compute an *F* ratio for the two sets of scores, we can predict that the *F* ratio computed for the three groups in *Part a* would probably be high and statistically significant, whereas the *F* ratio computed for the three groups in *Part* b would probably be lower and not statistically significant. This prediction is based on the knowledge that when the group variances are low (as is the case in *Part a*), MS_w (the denominator in the *F* ratio computation) is low, and we are more likely to get a high *F* ratio. The variances in *Part b* are higher than those in *Part a* and we can expect a higher MS_w and a lower *F* ratio.

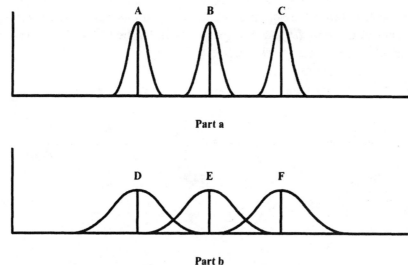

Figure 11.3   A graph showing three distributions with small variances (*Part a*) and a graph showing three
distributions with the same means as the groups in *Part a*, but with higher variances (*Part b*)

Once the *F* ratio is obtained, the *table of critical values for the F distribution* is consulted to determine whether to retain or reject the null hypothesis. *Retaining* the null hypothesis means that the sample means are not significantly different from each other beyond what might be expected purely by chance and we consider them as coming from the same population. *Rejecting* the null hypothesis means that at least two sample means differ significantly from each other.

In studies where the null hypothesis is rejected, the next step is to conduct a **post hoc comparison** in which all possible pairs of means are compared in order to find out which pair(s) of means differ(s). When the researcher predicts which means are expected to differ *before* starting the investigation, a method of *a priori* (or **planned**) **comparison** is used to test this prediction. *A priori* comparisons are appropriate when the researcher has a sound basis for predicting the outcomes before starting the study, while post hoc comparisons are appropriate in exploratory studies or when no specific prediction is made prior to the start of the study. In this book we demonstrate the computations of the *Tukey post hoc comparison* method using the numerical data in Table 11.2.

Hypotheses for a One-Way ANOVA

A one-way ANOVA tests the null hypothesis (H_o) that states that all the groups represent populations that have the same means. When there are three means, the *null* hypothesis is:[5]

$$H_0: \mu_1 = \mu_2 = \mu_3$$

The *alternative* hypothesis, H_A (also called the research hypothesis) states that there is a statistically significant difference between at least two means. When there are three groups, the alternative hypothesis is:

$$H_A: \mu_1 \neq \mu_2 \quad \text{and/or} \quad \mu_1 \neq \mu_3 \quad \text{and/or} \quad \mu_2 \neq \mu_3$$

The ANOVA Summary Table

The results of the ANOVA computations are often displayed in a summary table (see Table 11.1). This table lists the sum of squares (SS), degrees of freedom (*df*), mean squares (MS), *F* ratio (*F*), and the level of significance (*p* level). The general format of the ANOVA summary table is:

Table 11.1   The General Format of the One-Way ANOVA Summary Table

Source	SS	df	MS	F	p
Between	SS_B	$K-1$	MS_B	F-ratio	<.05>
Within	SS_W	N-K	MS_W		
Total	SS_T	N-1			

---

5.   **A HINT**: Although *samples* are studied, as with other statistical tests, we are interested in the *populations* that are represented by these samples. Therefore, in the null and alternative hypotheses, μ (the Greek letter *mu*) is used to represent the population means.

Instead of being displayed in a summary table, the results may also be incorporated into the text. The information in the text includes the F ratio, the degrees of freedom for the numerator (dfB) and the degrees of freedom for the denominator (dfW). The information is listed as: $F_{(dfB, dfW)}$. The text may also include the level of statistical significance (p level).

Further Interpretation of the F Ratio

Figure 11.4, *Parts a, b*, and *c*, represent three hypothetical samples and their F ratios. *Part a* depicts production output from three shifts in the same manufacturing plant. Note that the distributions of the three shifts overlap a great deal and the means are not very different from each other. The F ratio comparing the three shifts is probably small and not statistically significant. *Part b* shows mean spending levels on luxury products for consumers in low, middle, and high income groups. As is expected, the low income group spends the least, and the high income group spends the most. The F ratio comparing the low and high income groups is most likely statistically significant. The figure also shows clear differences between the low and middle income groups, and between the middle and high income groups. Therefore, the ANOVA test comparing these groups to each other is likely to yield an F ratio that is statistically significant. *Part c* shows mean scores on an aggression scale, given to three groups after an intervention designed to decrease aggression amongst employees. The three groups are: control (C), placebo (P), and experimental (E). Note that after the intervention the experimental group had the lowest aggression mean score, followed by the placebo group, while the control group scored the highest. The difference between the experimental group and the placebo group may be statistically significant, and, quite likely, there is a statistically significant difference between the experimental and control groups. The difference between the placebo and the control groups is probably not statistically significant. We can speculate that in this hypothetical example, the F ratio is probably large enough to exceed the corresponding critical value, leading us to reject the null hypothesis.

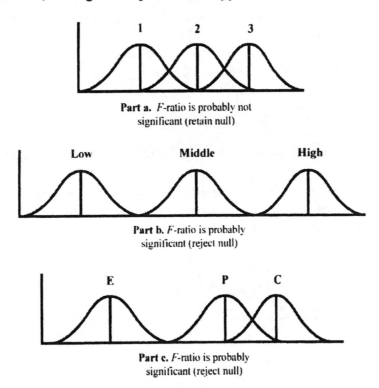

Part a. *F*-ratio is probably not
significant (retain null)

Part b. *F*-ratio is probably
significant (reject null)

Part c. *F*-ratio is probably
significant (reject null)

Figure 11.4   Three sets of distributions showing different F ratios: a nonsignificant F ratio (*Part a*);
a significant F ratio (*Part b*); and a significant F ratio (*Part c*)

An Example of a One-Way ANOVA

Professor Learner, the statistics course instructor at Midwestern State University, wants to test four instructional methods for teaching statistics. Students who signed up to take her statistics course are assigned at random to four sections: Section 1 is taught online using the Internet to deliver the course; Section 2 is taught using lectures; Section 3 is taught using independent study; and Section 4 is taught using a combination of lectures, group work, and online work. In this study, the instructional methods are the independent variable. Students in all four sections have to take five quizzes and a comprehensive final examination (the dependent variable). The scores on each quiz can range from 1 to 15. With four sections, the null hypothesis (H_0) and the alternative hypothesis (H_A) are:

$$H_O : \mu_1 = \mu_2 = \mu_3 = \mu_4$$
$$H_A : \mu_i \neq \mu_j$$

The subscripts i and j can represent any two of the four sections. In other words, the alternative hypothesis predicts that there will be a significant difference between at least two of the four means. The null hypothesis predicts that there would be no significant differences between the section means beyond what might happen purely by chance, due to some sampling error.

To illustrate the computations of ANOVA, we choose at random the scores of five students from each section on one of the quizzes. These quiz scores are the dependent variable. Of course, if this was a real study, we would have used a much larger sample size!

Table 11.2 lists the scores of the 20 randomly selected students. The sample sizes and means $(n$ and $\overline{X})$ are listed for each section and for the total group $(N$ and $X_T)$. As we can see, there are differences between the means of the four groups $(X_1=13.8, X_2=13.2, X_3=11.2,$ and $X_4=14.6)$. The question is whether these differences are statistically significant or due to chance and ANOVA can help us answer this question. (Note that our numerical example does not include the computational steps because we are likely to use a computer program to do the calculations for us.)

Table 11.2   Raw Scores and Summary Scores of Four Groups on a Statistics Quiz

Online Section 1	Lectures Section 2	Independent Study Section 3	Combined Section 4	TOTAL
14	14	11	15	
15	13	10	14	
13	11	11	15	
13	13	14	14	
14	15	10	15	
$n_1=5$	$n_2=5$	$n_3=5$	$n_4=5$	$N_T=20$
$\overline{X}_1=13.80$	$\overline{X}_2=13.20$	$\overline{X}_3=11.20$	$\overline{X}_4=14.60$	$\overline{X}_T=13.20$

The numerical results are displayed in an ANOVA summary table (Table 11.3). The table lists the three sources of variability (SS), the three degrees of freedom (df), the two mean squares (MS), the F ratio, and the p value.

Table 11.3   **One-Way ANOVA Summary Table for the Data in Table 11.2**

Source	SS	df	MS	F	p
Between	31.6	3	10.53	7.14	<.01
Within	23.5	16	1.48		
Total	55.2	19			

Even though the p value is given in the ANOVA summary table, let's follow the process for determining whether our obtained F ratio of 7.14 is statistically significant. The partial F distribution table in Figure 11.5 is used to find the appropriate critical values. In most textbooks, the table for finding the significance level of F ratios includes the p values at the .05 and .01 levels only. In Figure 11.5 the numbers on the lower row represent the .01 level of critical values and the numbers just above it are the .05 critical values. The numbers in the columns are the degrees of freedom *between* (*K-1*, or the number of groups minus 1) and the numbers in the rows are the degrees of freedom *within* (*N-K*, or the total number of scores minus the number of groups). In our case, df_B is 3 (4 groups minus 1) and df_w is 16 (20 students minus 4 groups). The critical F value at p = .05 level is 3.24; and at the p = .01 level, the critical value is 5.29. These critical values can be listed as:

$$F_{crit\,(.05,3,16)} = 3.24 \qquad F_{crit\,(.01,3,16)} = 5.29$$

The first number in the parentheses is the p value. The second number is the *df* for *between* (the numerator, or the column). And, the third number is the *df* for *within* (the denominator, or the row).

df within	*p* value	Degrees of Freedom Between		
		2	3	4
16	.05	3.63	3.24	3.01
	.01	6.23	5.29	4.77

Figure 11.5   Partial distribution of F values

The obtained F ratio of 7.14 *exceeds* the critical F value at p=.05 and at p=.01 probability levels and we report the results as significant at p<.01. The significant F ratio indicates that there is a statistically significant difference between at least one pair of means and the null hypothesis (H_0) is rejected in favor of the alternative hypothesis (H_A). In making the decision to reject the null hypothesis, there is less than 1% chance of making a Type I error (i.e., rejecting the null hypothesis when, in fact, it should be retained).

Our conclusion is that the instructional method *does make a difference* in the students' test scores. As is indicated in Table 11.2, students in Section 4 where the instructional method was a combination of lectures, online, and group work, obtained the highest quiz scores. The second highest mean quiz score was obtained by students in Section 1 (Online), followed by Section 2 (Lectures). The students in Section 3 (Independent Study) obtained the lowest mean score on the quiz.

Since our F ratio was significant at the p<.01 level, our next step is to conduct a *post hoc comparison* to find out which means are significantly different from each other. We use the Tukey method for the post hoc comparison.

Post Hoc Comparisons

The **Tukey method** of *post hoc multiple comparisons* is also called the **honestly significant difference (HSD)**. The HSD value tells us what the difference between a pair of means should be in order to view the means as significantly different from each other. Any difference that exceeds the HSD value is considered statistically significant.

When analyzing your numerical data, you are most likely to use a computer program. Although computer statistical programs can provide the HSD values for you, we present here an explanation of the computations of the HSD values using our numerical example. To simplify the computations, we will use the Tukey's HSD method for equal group size. The formula for the HSD value is:

$$HSD = Q_{(dfW,K)}\sqrt{\frac{MS_W}{n}}$$

Where		
Q	=	Value obtained from the Studentized Range Statistic table
df_w	=	Within groups degrees of freedom
K	=	Number of groups (or means)
MS_w	=	Within groups mean squares
n	=	Number of people in a group

The first step is to consult a table called the **Studentized Range Statistic** (part of which is reproduced in Figure 11.6 below). The table contains Q **values** for the .05 and .01 levels of significance. The rows list the degrees of freedom within (df_w) and the columns list the number of groups (K). In our example, df_w are 16, and K is 4. Therefore, the corresponding Q value for the .05 level of significance is 4.05, and for the .01 level it is 5.19.

df for within mean square	*p* value	k = number of groups								
		2	3	4	5	6	7	8	9	10
16	.05	3.00	3.65	4.05	4.33	4.56	4.74	4.90	5.03	5.15
	.01	4.13	4.78	5.19	5.49	5.72	5.92	6.08	6.22	6.35

Figure 11.6 Partial Distribution of Studentized Range Statistic

According to Table 11.3, MS_w (*MS* within) is 1.48 and n is 5. The computations of the HSD value at the p=.05 and p=.01 levels of significance are:

$$HSD_{(.05)} = 4.05_{(16,4))}\sqrt{\frac{1.48}{5}} = 4.05\sqrt{0.295} = 4.05(0.54) = 2.19$$

$$HSD_{(.01)} = 5.19_{(16,4))}\sqrt{\frac{1.48}{5}} = 5.19\sqrt{0.295} = 5.19(0.54) = 2.80$$

Once we find the HSD value, we set up a table in which we list all the means in *ascending* order, and subtract the *lower* mean from the *higher* mean at each intersection of two means. (See Table 11.4.) For example, we subtract the mean of Section 3 (11.20) from the mean of Section 2 (13.20) and record the difference (2.00) in the post hoc table (Table 11.4). We then compare the difference between all pairs to the HSD values at p<.05 and p<.01. These HSD values are 2.19 and 2.80, respectively.

Table 11.4   Mean Scores of Four Groups on a Statistics Quiz: Post Hoc Comparison

	Independent Study $\bar{X}_3=11.20$	Lectures $\bar{X}_2=13.20$	Online $\bar{X}_1=13.80$	Combined $\bar{X}_4=14.60$
$\bar{X}_3 = 11.20$	--	2.00	2.60*	3.40**
$\bar{X}_2 = 13.20$		--	0.60	1.40
$\bar{X}_1 = 13.80$			--	0.80
$\bar{X}_4 = 14.60$				--

*$p<.05$ **$p<.01$

According to Table 11.4 , the difference between the means of Sections 1 and 3 is significant at the $p<.05$ level, and the difference between the means of Sections 3 and 4 is significant at the $p<.01$ level. No other means are significantly different from each other. Figure 11.7 depicts the same four groups:

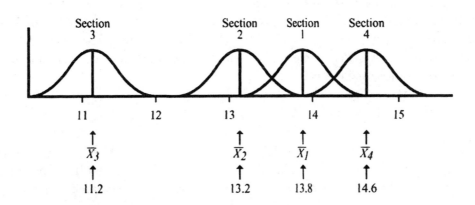

Figure 11.7   A graph illustrating the data in Table 11.4

According to Figure 11.7, students in Section 3 (Independent Study) scored the lowest and students in Section 4 (Combined) scored the highest. The difference between these two groups was significant at $p<.01$. The difference between the students in Section 3 (Independent Study) and Section 1 (Online) was also significant, but at a lower level ($p<.05$). The other pair-wise comparisons were not statistically significant. The likelihood that the decision to reject the null hypothesis is wrong is low and Professor Learner can be quite confident that her choice of instructional method in teaching statistics affects the students' quiz scores. The professor can examine the results in order to decide which instructional methods to use in teaching the statistics course in the future.

Two-Way ANOVA

Conceptualizing the Two-Way ANOVA

The *two-way* ANOVA test is designed to study the relationship between *two* independent variables and a dependent variable. One advantage of the two-way ANOVA is that it can reveal an *interaction* between the two independent variables. This interaction may not be apparent when a series of one-way ANOVA tests is conducted. To illustrate this point, let's look at the example that was used to demonstrate the computation of a one-way ANOVA.

Four different instructional methods were tested in four sections of a college statistics course. Students who registered to take the course were assigned at random to one of the four sections.

The means on a quiz administered to students in all four sections were compared to each other, using a one-way ANOVA. Suppose we want to further divide the students in each group by their major in college, by ability level, or by gender. It is possible to conduct another ANOVA test to compare, for example, the quiz scores of Accounting students in all four instructional methods to the scores of their classmates who major in Management. We can run two separate one-way ANOVA tests; one to compare the four methods and one to compare the two majors. However, instead of running these two tests, we can do a single two-way ANOVA test. The two-way ANOVA would allow us to compare simultaneously the method effect, the college major effect, and the possible effect of the interaction between the method and the major on the students' quiz scores. For example, Accounting students may score higher using one instructional method, whereas Management students may do better using another method.

The total variation in a two-way ANOVA is partitioned into two main sources: the *within-groups* variation and the *between-groups* variation. The between group variation is further partitioned into three components: (a) the variation among the *row means*, (b) the variation among the *column means*, and (c) the variation due to the *interaction*.

Four mean squares (MS) are computed in a two-way ANOVA. Two are computed for the *two main effects* (the independent variables), one is computed for the *interaction*, and one for the *within*. Then, using the mean squares within (MS_w) as the denominator, three F ratios are computed. These F ratios are found by dividing each of the three mean squares (MS_{Row}, MS_{Column}, and $MS_{Row \, X \, Column}$) by MS_w. As was the case with a one-way ANOVA, a summary table is used to display the two-way ANOVA summary information. The table includes the sum of squares, degrees of freedom, mean squares, F ratios, and p levels.

Hypotheses for the Two-Way ANOVA

A two-way ANOVA is conducted to test three null hypotheses about the effect of each of the two independent variables on the dependent variable and about the interaction between the two independent variables. The two independent variables (or *factors*) are referred to as the *row* variable and the *column* variable.

To test the three null hypotheses, three F ratios are calculated in a two-way ANOVA. The three null hypotheses are:

$H_{O(Row)}$:   There are no statistically significant differences among population means on the dependent measure for the first factor (the *row factor*).

$H_{O(Column)}$:   There are no statistically significant differences among the population means on the dependent measure for the second factor (the *column factor*).

$H_{O(Interaction)}$:   In the population, there is no statistically significant interaction between Factor 1 and Factor 2 (the *row x column interaction*).

Graphing the Interaction

It is often helpful to further study the interaction by graphing it. To create *the interaction graph*, the mean scores on the dependent variable are marked on the vertical axis. Lines are then used to connect the means of the groups. Suppose, for example, we want to conduct a study to investigate the effectiveness of two advertisements designed to increase purchase intentions of consumers who have already tried the product as well as consumers who have not tried the product. In this study, whether the consumer has tried the product before is one independent variable and the second independent variable is the two different advertisements. The dependent variable is the purchase intention. Half of the consumers who have tried the product and half of who have not tried the product watch Advertisement 1. The other half in each group of consumers who have and have not seen tried the product watch Advertisement 2. Figure 11.6 shows two possible outcomes of the study: The interaction is significant and the lines intersect (*Part a*); and the interaction is not significant and the lines are parallel (*Part b*).

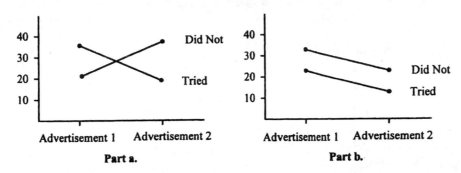

Figure 11.8   A graph showing a statistically significant interaction (*Part a*) and a graph showing an interaction that is not statistically significant (*Part b*)

Part a in Figure 11.8 shows that Advertisement 1 was more effective with the consumers who had tried the product and Advertisement 2 was more effective with consumers who had *not* tried the product. This is an example of an interaction effect. *Part b* shows no interaction effect. Advertisement 1 was more effective for both consumers who had and had *not* tried the product, and Advertisement 2 was less effective for both groups of consumers. In addition, the consumers who already tried the product and saw either Advertisement 1 or Advertisement 2 had lower purchase intention scores than the consumers who did *not* try the product and saw Advertisement 1 or Advertisement 2.

The interaction may be significant even if the two lines do not intersect, as long as they are *not parallel*. To illustrate this point, let's look at the data in Table 11.5. In this hypothetical example, four groups of customer service representatives working for two different managers are trained to manage the stress associated with their jobs using two different stress management methods. One *independent* variable is the customer service representatives' manager (Manager A and Manager B) and the other *independent* variable is the stress management method (Method 1 and Method 2). The *dependent* variable is the customer service representatives' score on a stress management. One group of customer service representatives that work for each of the two managers is practicing stress management Method 1 and the other group of customer service representatives that work for each manager is practicing stress management Method 2.

Table 11.5   **Means on a Stress Management Test of Two Groups
Using Two Different Stress Management Methods**

	Method 1	Method 2
Manager A	20	15
Manager B	25	35

Table 11.5 shows the mean scores of the four groups. An inspection of the four means indicates that customer service

representatives working for Manager A who were using Method 1 scored higher on the stress management test compared with their peers working for the same manager who were using Method 2. For Manager B, the results were reversed. In this case, customer service representatives who were using Method 2 scored higher than their peers who were using Method 1. The table also shows that the customer service representatives working for Manager B scored higher than the customer service representatives working for Manager A.

Figure 11.9 shows the interaction effect of the data displayed in Table 11.5. Although the two lines representing the two groups do not cross, they are on a "collision course," which is typical of a significant interaction effect.

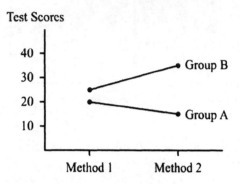

Figure 11.9   A graph of the data in Table 11.5 showing a significant interaction

There are two types of significant interactions: (a) **disordinal**, where the lines intersect; and (b) **ordinal**, where the lines do not intersect (but are not parallel). *Part a* in Figure 11.8 shows an example of a *disordinal* interaction and Figure 11.9 depicts an *ordinal* interaction.

When, in addition to having a significant interaction, the two main effects (the row variable and the column variable) are also significant, it may be difficult to interpret the results. Figure 11.10 illustrates another hypothetical example of interaction, showing the means of two managers' groups of customer service representatives (Manager A and Manager B) and two stress management methods (Method 1 and Method 2). As you can see, the interaction is significant (the lines cross). Manager A's group scored higher than Manager B's group when Method 2 was used, and Manager B's group scored a bit higher than Manager A's group when Method 1 was used.

Looking at Figure 11.10, we can speculate that in addition to a significant F ratio for the interaction, the two F tests for main effects (methods and groups) are also significant. Those interpreting the results from the study should exercise caution when they make a decision about the efficacy of the two managers and the two stress management methods employed by the company.

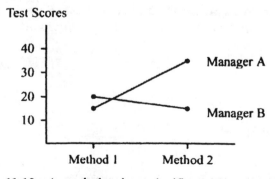

Figure 11.10   A graph showing a significant interaction and significant main effects

The Two-Way ANOVA Summary Table

As was mentioned, the results of the computations of a two-way ANOVA test are presented in a summary table, similar to the table that is used for presenting the results of a one-way ANOVA (see Table 11.1). Each of the two factors in a two-way ANOVA (which are also called the **main effects**) is associated with an F ratio. Similarly, the interaction is analyzed using its own F ratio. As in the one-way ANOVA, a summary table is used to display the results of the two-way ANOVA analyses (see Table 11.6). Note, though, that there are three F ratios in Table 11.6, as opposed to a single F ratio in the one-way ANOVA (see Table 11.1). The three F ratios in a two-way ANOVA are designed to test the three null hypotheses. Table 11.6 also shows the computations of the degrees of freedom associated with each of the three F tests.

Table 11.6		Two-Way ANOVA Summary Table			
Source	SS	df	MS	F	p
Main Effects					
Factor 1 (row)	SS_R	no. of levels $-$ 1	MS_R	F_R	<.05>
Factor 2 (column)	SS_C	no. of levels $-$ 1	MS_C	F_C	<.05>
Interaction	SS_{RxC}	df_{row} x $df_{col.}$	MS_{RxC}	F_{RxC}	<.05>
Within Groups	SS_W	$N-K$	MS_W		
Total	SS_T	$N-1$			

An Example of a Two-Way ANOVA

A Human Resources Manager trying to attract more female management trainees wants to know if the gender of the upper level executives in a company makes a difference in business school students' interest in working for that company. The manager conducts a survey of 56 business students in their last semester of undergraduate school. The sample includes 28 females and 28 males.[6] Fourteen of the female students and 14 of the male students listened to a wide variety of information about a company, including the fact that the majority of the upper level executives in that company are female. The other 14 female students and 14 male students listened to the same types of information provided about the company, only this time they were told the majority of the upper level executives are male. Except for the gender of the top executives, the four companies are otherwise similar in terms of sales, profitability, number of employees, and industries in which they operate. After listening to the information on the companies the students completed an interest inventory designed to measure their interest in the companies about which they received information. Scores on the interest inventory can range from 10 to 20 points. Table 11.7 presents the scores of all 56 students, as well as the groups' means. The rows show the scores of the male and female students and the columns show the scores of the groups of students who were exposed to the two types of companies (female-led companies and male-led companies).[7]

---

6. **A HINT**: Although ANOVA assumes that the groups are random samples from their respective populations, in many studies this assumption may not be fully met. As mentioned before, empirical studies have shown that ANOVA can be conducted under such conditions without seriously affecting the results, *especially when the group sizes are similar.*

7. **A HINT**: Remember that in ANOVA, each mean represents a separate and independent group of individuals and no person can appear in more than one group. In our example, each student was given information about a company where the majority of the upper level executives were female *or* a company where the majority of the upper level executives were male (but *not both types* of companies).

Table 11.7   Scores and Means of Students' Interests of
Male Versus Female Led Companies

	Type of Company		
	Female Upper Level Executives	Male Upper Level Executives	Total
Male Students	14 13 17 10 18 16 15 15 15 16 14 17 16 15 $\bar{X}$=15.07	19 17 18 20 18 17 16 18 20 19 18 17 16 17 $\bar{X}$=17.86	$\bar{X}_M$=16.46
Female Students	17 20 19 19 20 18 16 15 17 19 18 16 17 18 $\bar{X}$=17.79	18 14 16 15 14 13 17 15 14 14 13 16 15 16 $\bar{X}$=15.00	$\bar{X}_F$=16.39
Total	$\bar{X}_{FE}$=16.43	$\bar{X}_{ME}$=16.43	

Note that the total mean of the 28 male students on the interest inventory (X_M=16.46) is almost identical to the total mean of the 28 female students (X_F=16.39). There is also no difference in the students' interest in whether men or women lead the company (see the two column totals). As we can see, the mean score for the male students and female students who were given information about companies with female executives (X_{FE}=16.43) is identical to the mean score for the group of male students and female students given information about companies led by mostly males (X_{ME}=16.43).

The interaction of student gender (the row factor) and the type of company (the column factor) is displayed in Figure 11.11. The figure shows that the lines cross, indicating a significant interaction of the two factors. There is a clear higher level of interest by male students in the company whose high level executives are males while female students are more interested in the company led by female executives.

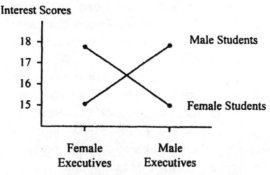

Figure 11.11   A graph showing the mean scores and significant interaction of the two factors (student gender and gender of the majority of the executives) of the data in Table 11.7

The two-way ANOVA summary table (Table 11.8) lists the sum of squares (SS), the degrees of freedom (*df*), the mean squares (MS), the *F* ratios, and the level of significance (*p* value).[8] The *F* ratios in the summary table confirm

8.   **A HINT**: The detailed computations of the three *F* rations are not included in this book. The data presented in Table 11.8 were analyzed using a computer statistical software package.

our earlier inspection of Table 11.7. The F ratio for rows (comparing the interests of male and female students) and the F ratio for columns (comparing the interest in the two types of companies) are not statistically significant ($F=0.03$ and $F=0.00$, respectively). However, the F ratio for the interaction effect is very high ($F=43.13$) with a p value of .001. We can be very confident about our decision to reject the null hypothesis. There is a strong interaction between the students' gender and their interest in joining a company where the top-level executives are of their same gender.

Table 11.8 **Two-Way ANOVA Summary Table**

Source	SS	df	MS	F
Main Effects				
Gender	0.07	1	0.07	0.03
Type of Company	0.00	1	0.00	0.00
Interaction	108.64	1	108.64	43.13*
Within[9]	131.00	52	2.52	
Total	239.71	55	4.36	

*$p<.001$

If we had conducted two one-way ANOVA tests instead of the single two-way ANOVA test, we might have concluded the following: (a) there is no difference in the level of interest expressed by male students and female students toward the two types of companies (a nonsignificant F ratio for rows), and (b) the gender of the majority of the executives in the company does *not* make a difference in student's attitudes toward the companies (a nonsignificant F ratio for the columns). However, upon inspecting the means in Table 11.7, the interaction diagram in Figure 11.10, and the level of statistical significance (p values) in Table 11.8, it becomes clear that there is a difference between male and female students in their level of interest toward the two types of companies. In our fictitious study, the male students were more interested in companies with a majority of males as upper-level executives, and the female students were more interested in companies with a majority of females as upper-level executives. If this were a real study, instead of a fictitious one, the implications for the Human Resource managers would have been that if their organizations are led by mostly females it should make it easier for them to attract female students to the companies. Similarly, male students would be more attracted to companies where the top executives are males.

Summary

1. An **analysis of variance (ANOVA)** is used to compare the means of two or more independent samples and to test whether the differences between the sample means are statistically significant.

2. The **one-way analysis of variance (one-way ANOVA)** can be thought of as an extension of a t test for independent samples. It is used when there are two or more independent groups.

3. The *independent* variable is the categorical variable that defines the groups that are compared. The *dependent* variable is the measured variable whose means are being compared.

---

9.   **A HINT**: The "Within" (i.e., within groups) listed under the "Source" column is also called **Residual** or **Error**.

4. There are several assumptions for ANOVA: (a) the groups are independent of each other, (b) the dependent variable is measured on an interval or ratio scale, (c) the dependent variable is normally distributed in the population, (d) the scores are random samples from their respective populations, and (e) the variances of the populations from which the samples were drawn are equal (the assumption of the *homogeneity of variances*). The first two assumptions cannot be violated.

5. ANOVA is considered a *robust* statistic that can stand some violation of the third and fourth assumptions. Empirical studies show that there are no serious negative consequences if these assumptions are not fully met. The assumption of the homogeneity of variances can be tested using special tests.

6. The test statistic in ANOVA is called the **F statistic**. The **F value** (or **F ratio**) is computed by dividing two variance estimates by each other. If the F ratio is statistically significant (i.e., if $p < .05$) and if there are three or more groups in the study, then a pair-wise comparison is done to assess which means are significantly different from each other.

7. When one independent variable is used, the design is called a *one-way ANOVA*; when two independent variables are used, the design is called a **two-way ANOVA**. In general, when two or more independent variables are used, the design is called a **factorial ANOVA**.

8. In the factorial ANOVA test, the **interaction** refers to a situation where one or more levels of the independent variable have a different effect on the dependent variable when combined with another independent variable.

9. An independent variable must have at least two *levels* (or conditions). For example, the variable of gender has two levels (female and male) and the variable of the seasons of the year has four levels (fall, winter, spring, and summer).

10. The *null hypothesis* (H_o) in a one-way ANOVA states that there is no significant difference between the population means; the *alternative hypothesis* (H_A) states that at least two population means differ significantly from each other.

11. The *variability* in a one-way ANOVA is divided into three *sums of squares (SS)*: **within-groups sum of squares (SS_W)**, **between-groups sum of squares (SS_B)**, and **total sum of square (SS_T)**. The SS_T is equal to the combined SS_W and SS_B.

12. The variance estimates in ANOVA are called the *mean squares*. The **mean squares between (MS_B)** and the **mean squares within (MS_W)** are obtained by dividing SS_B and SS_W by their corresponding degrees of freedom.

13. The MS_W (also called the **error term**) can be thought of as the average variance to be expected in any normally distributed group. The MS_W serves as the denominator in the computation of the F ratio.

14. The F ratio is obtained by dividing the MS_B by the MS_W:

$$F = \frac{MS_B}{MS_W}$$

15. The obtained F ratio is compared to the appropriate critical values using the *table of critical values* for the F statistic.

16. The results of the ANOVA test are often displayed in a **summary table**. The summary table includes the sum of squares, degrees of freedom, mean squares, F ratio, and level of significance (*p* value).

17. If the F ratio is statistically significant ($p<.05$), a **post hoc comparison** test is conducted to determine which means are significantly different from each other. An example of a post hoc comparison procedure, called the **Tukey's honestly significantly difference (HSD)** test, is demonstrated in the book. The HSD value tells us what the difference between a pair of means should be in order to conclude that the means are significantly different from each other.

18. The *two-way ANOVA* test is used to compare two independent variables (or factors) *simultaneously*. This statistical test enables us to study the effect of each of the two factors on the dependent variable as well as the **interaction** of the two factors. The independent variables in factorial ANOVA are also called the **main effects**.

19. A two-way ANOVA is conducted to test three hypotheses about differences between the *row* variable, the *column* variable, and the *interaction* of these two independent variables (or factors). Three F ratios are calculated to test each of the three null hypotheses.

20. The *total* variation in a two-way ANOVA is partitioned into two main sources: The *within-groups* variation and the *between-groups* variation. The between-groups variation is further partitioned into three components: The variation among the *row mean*, the variation among the *column means*, and the variation due to *interaction*.

21. As was the case with a one-way ANOVA, the two-way ANOVA summary information is presented in a table. The summary table includes four sums of squares, four degrees of freedom, three mean squares, three F ratios, and three p levels.

22. In order to better understand an interaction, it is often helpful to graph it. To create the *interaction graph*, the mean scores on the dependent variable are marked on the vertical axis. Lines are then drawn to connect the means of the groups.

23. A *nonsignificant* interaction is represented by *parallel* lines and a *significant* interaction is represented by *nonparallel* lines.

24. There are two types of significant interactions: (a) **disordinal**, where the lines intersect; and (b) **ordinal**, where the lines do not intersect (but are not parallel).

Chapter 12

Chi Square Test

The **chi square test**, represented by χ^2, is applied to discrete data (i.e., nominal, categorical data). The units of measurements that are often used are frequency counts and observations (rather than scores). The chi square test was developed by Karl Pearson (who also developed the Pearson product moment correlation) in 1900, in order to measure how well observed data fit a theoretical distribution. The chi square test belongs to a group of statistical methods called **nonparametric**, or distribution-free. These statistical tests can be applied to data that do not meet certain assumptions (e.g., being measured on an interval or ratio scale, or being normally distributed).

The chi square statistic can be used to analyze data measured on a *nominal* scale, such as gender, where there are two or more discrete categories. It can also be used to analyze other types of numerical data (such as data measured on an interval scale) that are first divided into *logically defined* and generally agreed upon categories. For example, IQ scores can be divided into three categories (high, average, and low) by using the standard deviation of the IQ scale to define each category. (See Chapter 2 for a discussion of measurement scales.)

The chi square test is often used to analyze questionnaire data where a numerical code is assigned to groups or responses. For example, the political affiliations of registered voters may be assigned a numerical code of 1-4 as follows: *Democrats*=1; *Republicans*=2; *Independents*=3; and *Other*=4.[1] In another example, a number is assigned to each position in a company: "Accountant" is coded as 1; "Engineer" is coded as 2, and "Support Staff" is coded as 3.

In applying the chi square test, two types of frequencies are used: *observed* and *expected*. The **observed frequencies (O)** are based on actual (empirical) observations and on "head counts." An example of observed frequencies is the actual number of people who respond "*Yes*" or "*No*" to a particular question. The **expected frequencies (E)** are theoretical or based on prior knowledge. The observed and expected frequencies can be expressed as actual head counts or as percentages. (The process for determining the expected frequencies is explained in this chapter.)

The chi square test is used to determine whether there is a significant difference between the observed and expected frequencies and both types of frequencies are used in the computation of the **chi square value (χ^2)**. Each pair of observed frequencies and its corresponding expected frequencies is called a **cell**. To compute χ^2, for each cell we start by computing $(O-E)^2$ then dividing it by E. We then add up the results of the computations from each cell to obtain the χ^2 value. The formula for χ^2 is:

$$\chi^2 = \Sigma \left[\frac{(O-E)^2}{E} \right]$$

---

1.   **A HINT**: Political affiliation is a nominal scale variable and the code we assigned in this example was arbitrarily chosen (see Chapter 2).

Where χ^2 = Chi square statistic
              O   = Observed frequencies for each cell
              E   = Expected frequencies for each cell

The numerator in the equation is the difference between the *observed* and *expected* frequencies. When there are very small differences between the observed and expected frequencies in each cell, the numerator is small and the chi square value is low. On the other hand, when there are large differences between the observed and expected frequencies in each cell, the numerator is large, and, in turn, so is the chi square value.

The degrees of freedom (*df*) in the chi square statistic are related to the number of *levels* (i.e., categories) in the dependent variable(s). This is different from the procedures for computing the degrees of freedom in other statistical tests, such as Pearson correlation and *t* test, where the degrees of freedom are related to the sample sizes. This chapter includes explanations of the procedures for calculating the degrees of freedom used in the chi square analysis.

Assumptions for the Chi Square Test

Various types of data can be analyzed using the chi square statistic. However, several assumptions are required in order to apply the chi square test. These assumptions are:

1.   The observations should be *independent* of each other and a particular response cannot be counted in more than one category. For example, a person may not be asked on two different occasions to respond to the same question, as if two people each responded once. The total number of observed frequencies should not exceed the number of participants.

2.   The data must be in the form of *frequencies*. The total number of *observed* frequencies must equal the total number of *expected* frequencies.

3.   The categories, especially those that comprise ordinal, interval, or ratio scale, should be created in some *logical*, defensible way. The criteria used to establish the categories should be chosen carefully and wisely. For example, suppose one of the variables in a study is the participants' level of income. Income is a continuous variable and it is necessary to establish logical cutoff points to define the various categories. The researcher may want to divide the variable of *income* into categories (e.g., *High*, *Middle*, and *Low*) by following some official guidelines for defining income levels.

There are two types of chi square tests: The chi square **goodness of fit test**, which is used with one independent variable; and the chi square **test of independence**, which is used with two independent variables. The chi square *goodness of fit test* is used to test the fit between a distribution of observed frequencies and a distribution of expected frequencies. The chi square *test of independence* is used to test whether two factors (independent variables) are independent of each other.

The Goodness of Fit Chi Square Test

In a goodness of fit chi square test, the number of expected frequencies in each category may be *equal* or *unequal*. The following is a discussion of the two types.

Equal Expected Frequencies

In this type of chi square test, there are *equal expected frequencies* in each category. The observed frequencies, as always, are based on empirical data—that is, on observation. We collect data by recording the number of occurrences in each category. For example, we can use the chi square procedure to test whether a coin is fair by repeatedly tossing the coin and recording the number of heads and tails. These numbers are our *observed* frequencies. The *expected* frequencies are based on the assumption that the coin is fair; thus, half of the time it should land heads, and half of the time, tails. The null hypothesis states that the coin is fair and, consequently, would land as many times heads as tails.

Suppose we toss a coin 100 times, and record 55 heads and 45 tails (see Table 12.1). With a fair coin (the null hypothesis is true), we would expect 50 heads and 50 tails. Therefore, we must ask ourselves whether our coin is fair or biased. As with other statistical tests, such as the *t* test, we cannot simply eyeball our observed data and decide whether the coin is biased based on our observations. In other words, it would be difficult to look at the difference between 55 and 45 and determine whether it is large enough to indicate a biased coin, or whether this difference is small enough to have occurred purely by chance.

Table 12.1   Observed and Equal Expected Frequencies for Heads and Tails

	O (Observed)	E (Expected)	$\dfrac{(O-E)^2}{E}$
HEADS	55	50	$\dfrac{(55-50)^2}{50} = \dfrac{25}{50} = 0.50$
TAILS	45	50	$\dfrac{(45-50)^2}{50} = \dfrac{25}{50} = 0.50$
TOTAL	100	100	$\chi^2 = 1.00$

The computations that are displayed in Table 12.1 show that we obtained a chi square value of 1.00. We now need to determine whether this test statistic is statistically significant by consulting the table of critical values for the chi square statistic. (See Figure 12.1 for a partial table of the distribution of chi square.) The degrees of freedom (*df*) are the number of categories, or levels, minus 1. In our example, the variable has two levels, heads and tails, resulting in a *df* of 1 (2-1=1). As before, unless told otherwise, we use the convention of 95% confidence level and inspect the critical values listed in the column of *p*=.05. With 1 degree of freedom (*df*=1), the *critical* value from the table of critical values is 3.841 (see Figure 12.1). Our *obtained* χ^2 value of 1.00 *does not* exceed this critical value; therefore, we *retain* the null hypothesis. Our conclusion is that the coin is fair even though when we tossed it 100 times it landed more times heads than tails. The difference between the heads and the tails is small enough to have happened by chance alone, and is probably due to a random error, rather than a systematic error (i.e., a biased coin).

			p value		
df	.10	.05	.02	.01	.001
1	2.706	3.841	5.412	6.635	10.827

Figure 12.1   Partial Table of Distribution of Chi Square (X^2)

Unequal Expected Frequencies

The chi square test for *unequal expected frequencies* is used primarily to compare similarities and differences between a group's observed frequencies and expected frequencies that are unequal. Often, the expected frequencies represent the population distribution of the observations on the variable being investigated. In the study, the researcher studies the match between the sample distribution (the observed frequencies) and the population distribution (the expected frequencies). The researcher has to know the expected frequencies *a priori* (ahead of time) in order to conduct this type of chi square analysis. An example may help to illustrate this chi square test.

A series of articles published in the local press reports that there is an ongoing problem of grade inflation in the School of Business at the State University. Reporters contend that too many grades of *A* and *B* are given to undeserving students at the School of Business. The dean of the school argues that the distribution of grades given to students in the School is similar to the distribution of grades in other similar institutions. The chi square statistic is selected to analyze the data and compare the distribution of grades in the School of Business (the observed frequencies) with the grades in other similar institutions (the expected frequencies) (Table 12.2). The null hypothesis states that there is no difference in the distribution of grades between the School of Business and other similar institutions. If the two distributions of observed and expected frequencies turn out to be similar, the resulting chi square value would be small, thus leading the researchers to retain the null hypothesis. As was discussed before, in order to be able to conduct this type of chi square test, the researcher has to have *a priori* knowledge about the distribution of the expected frequencies.

**Table 12.2   Observed and Unequal Expected Frequencies
for Five Letter Grades**

Grade	O	E	$\dfrac{(O-E)^2}{E}$
A	16	10	$\dfrac{(16-10)^2}{10} = 3.60$
B	22	20	$\dfrac{(22-20)^2}{20} = 0.20$
C	38	40	$\dfrac{(38-40)^2}{40} = 0.10$
D	17	20	$\dfrac{(17-20)^2}{10} = 0.45$
F	7	10	$\dfrac{(7-10)^2}{10} = 0.90$
TOTAL	100	100	$\chi^2 = 5.25$

The computations that are displayed in Table 12.2 show that the obtained chi square value is 5.25. The degrees of freedom are 4 (df=5-1=4). According to the table of critical values (Figure 12.2) the critical value of χ^2 at p=.05 and df of 4 is 9.488 ($\chi^2_{crit(.05,4)}$ = 9.488). Our obtained value of 5.25 *does not* exceed this critical value, leading us to *retain* the null hypothesis. Further inspection of the data shows that in comparison with other similar institutions, more grades of *A* and *B* and fewer grades of *D* and *F* were given in the School of Business. However, these differences are not statistically significant and do not indicate a great departure from the "norm." The School's dean may still want to review the grading process in the school, as they seem to differ somewhat from the standards in other colleges.

df	.10	.05	.02	.01	.001
4	7.779	9.488	11.668	13.277	18.465

p value

Figure 12.2   Partial Table of Distribution of Chi Square (χ^2)

The Chi Square Test of Independence

The chi square *test of independence* is conducted to test whether two independent variables (or factors) are related to, or are independent of, each other. For example, a researcher may want to investigate whether there is a difference in the political party affiliation between union members and managers in a company. The researcher may survey 100 union members and 100 managers, asking them to indicate whether they are Democrat, Republican, or Independent.[2] The responses of all participants are then tallied and arranged in a 2x3 ("two by three") table, as follows (Table 12.3):

Table 12.3   Observed Frequencies: Union and Managers Survey About Political Parties Affiliation

	Democrats	Republicans	Independents	TOTAL
Union	60	35	5	100
Managers	50	45	5	100
TOTAL	110	80	10	200

The null hypothesis states that political party affiliation is independent of group membership (i.e., there is no difference in the political affiliation distribution between the union members and the managers). The degrees of freedom for a two-variable chi square are calculated as the number of rows minus 1, multiplied by the number of columns minus 1: (Row - 1)(Column - 1). In the example above, there are 2 rows and 3 columns, so the degrees of freedom are 2 (*df*=[2-1]x[3-1]).

The most common tables in the chi square tests are those that have two levels in each of the two variables (e.g., males/females and yes/no). These are *contingency tables* that are referred to as *2x2* ("two by two") *tables*. Next, we use an example to take you through several of the steps in the computations of a chi square value using data presented in a 2x2 table. In this example, a group of accountants from 80 large companies and 90 small companies are asked to respond to the following question: "The company I work for pays its bills on time" by circling "yes" or "no." Table 12.4 displays their responses (the observed frequencies). The table also shows the row and column totals, as well as the **grand total**, which is the combined number of observed frequencies. You should be able to obtain the grand total either by adding up the row totals or by adding up the column totals.

Table 12.4   Observed Frequencies: Large and Small Company Accountants and Their Responses to the Question: "The Company I Work for Pays its Bills on Time"

	Yes	No	TOTAL
Large Company Accountants	(Cell A) 62	(Cell B) 18	80
Small Company Accountants	(Cell C) 57	(Cell D) 33	90
TOTAL	119	51	170 ⇑ Grand Total

2.   **A HINT**: When one of the variables is *group membership* (e.g., males and females), and the other variable is the responses of the *group members* (e.g., "yes" or "no"), the convention is to record the groups in the *rows* and their responses in the *columns*.

Table 12.5 displays the responses (the observed frequencies) as well as the expected frequencies for each of the four cells. However, the computational steps to obtain these expected frequencies are not included because they are tedious and readily available with computer statistical programs, such as SPSS. The values from each cell are then added to obtain a chi square value of 4.04. After obtaining the χ^2, we consult the table of critical values (Figure 12.3). Degrees of freedom are calculated as: (2-1)(2-1)=1. Consulting Figure 12.3 we can see that our obtained χ^2 value of 4.04 *exceeds* the critical value under *p* of .05 and 1 *df* ($\chi^2_{crit(.05,1)}$=3.81), but not the value under *p*=.02 and 1 *df* ($\chi^2_{crit(.02,1)}$=5.41). Therefore, we report the results of this chi square test to be significant at *p* <.05 and *reject* the null hypothesis. We are at least 95% confident that our decision to reject the null hypothesis is the right decision. The differences in the responses of the two groups of accountants are too large to have occurred purely by chance. We conclude that although in both groups the majority of accountants feel their companies pay their bills on time, the percentage of large company accountants responding positively to the question posed is higher than that of the smaller company accountants. The results of our study suggest that the accountants' responses may depend on the size of their companies.

As before, we use the observed and expected values in each cell to compute a cell value, using this formula:

Table 12.5   **Observed and Expected Frequencies of Large and Small Company Accountants and Their Responses to the Question: "The Company I Work for Pays its Bills on Time"**

Accountant Group	Yes		No		
	Observed	Expected	Observed	Expected	TOTAL
Large Company	62	56	18	24	80
Small Company	57	63	33	27	90

$$\frac{(O-E)^2}{E}$$

$$CellA: \frac{(62-56)^2}{56} = 0.64$$

$$CellB: \frac{(18-24)^2}{24} = 1.50$$

$$CellC: \frac{(57-63)^2}{63} = 0.57$$

$$CellD: \frac{(33-27)^2}{27} = 1.33$$

df		.10	.05	.02	.01	.001
p value	1	2.706	3.841	5.412	6.635	10.827

Figure 12.3   Partial Table of Distribution of Chi Square (χ^2)

Summary

1. The **chi square (χ^2) test** is applied to discrete, categorical data, where the units of measurement are frequency counts.

2. The chi square test is considered a nonparametric, or a distribution-free statistic. It can be used to analyze data measured on a *nominal* scale (such as gender) where there are two or more discrete categories. It can also be used to analyze other types of numerical data (such as data measured on an interval scale) that are first divided into *logically defined* and generally agreed upon categories.

3. The chi square test is often used to analyze questionnaire data where a numerical code is assigned to groups or responses.

4. In applying the chi square test, two types of frequencies are used: *observed* and *expected*. The **observed frequencies** are based on actual (empirical) observations and on "head counts." The **expected frequencies** are theoretical or based on prior knowledge. The observed and expected frequencies can be expressed as actual head counts or as percentages.

5. Each pair of observed frequencies and its corresponding expected frequencies is called a **cell**.

6. The observed and expected frequencies in each cell are used to compute the chi square value. In the formula for computing the chi square, "**O**" refers to the observed frequencies and "**E**" refers to the expected frequencies in each cell. The chi square value is found by adding up the results of the division (the quotient) from each cell.

$$\chi^2 = \sum \frac{(O-E)^2}{E}$$

7. The *degrees of freedom (df)* in the chi square statistic are related to the number of *levels* or *cells* (i.e., categories) in the independent variable(s). This is different from the procedures for computing the degrees of freedom in other statistical tests, such as Pearson correlation and *t* test, where the degrees of freedom are related to the sample sizes.

8. The assumptions required for applying the chi square statistic are: (a) the observations should be *independent* of each other, (b) the data are recorded as *frequencies* ("head count"), and (c) the categories are created in some *logical* and agreed upon way.

9. There are two types of chi square tests: (a) The **goodness of fit test**, with one independent variable, is used to test the fit between a distribution of observed frequencies and expected frequencies; and (b) The **test of independence**, with two independent variables, is used to test whether two factors (the independent variables) are independent of each other.

10. In a chi square *goodness of fit* test, the number of expected frequencies in each category may be *equal* or *unequal*. When equal expected frequencies are used, they represent the null hypothesis that posits that there is an equal probability of having the same number of frequencies in each cell. When unequal expected frequencies are used, they must be known *a priori* (ahead of time) and be provided by the researcher.

11. The chi square *test of independence* is used to test whether two variables are related to, or are independent of, each other. Each of the two variables has to have at least two levels (e.g., male/female, true/false, above/below).

12. The most common tables in the chi square tests are those that have two levels in each of the two variables (e.g., males/females and yes/no). These are contingency tables that are referred to as 2x2 ("two by two") tables.

13. When one of the variables is group membership (e.g., males and females), and the other variable is the responses of the group members (e.g., "yes" or "no"), the convention is to record the groups in the *rows* and their responses in the columns.

Part Six

Reliability and Validity

Chapter 13

Reliability

The term *reliable*, when used to describe a person, usually means that this person is dependable and consistent. Similarly, a reliable measure is expected to provide consistent and accurate results. If we use a reliable measure over and over again to measure physical traits, the same or very similar results should be obtained each time. For example, when we repeatedly use a precise scale to measure weight, we are likely to obtain the same weight time after time. However, when dealing with the affective domain (e.g., self-concept or motivation) or even with the cognitive domain (e.g., knowledge of a subject matter), the performance of individuals on a measuring tool tends to change and be much less consistent. Factors such as moods, pressure, fatigue, anxiety, and guessing all tend to affect performance. Therefore, even with a reliable measure, it may be hard to achieve a high level of consistency between measures. Since this book is intended for business practitioners, our discussion of reliability will focus on procedures used in business.

Reliability refers to the level of consistency of an instrument and the degree to which the same results are obtained when the instrument is used repeatedly with the same individuals or groups. This consistency may be determined by using the same measure twice, administering two equivalent forms of the measure, or using a series of items designed to measure similar concepts.

The symbol used to indicate the reliability level is *r*, the same as that used for the Pearson product-moment correlation coefficient (see Chapter 8). As will be explained later in this chapter, several procedures to assess reliability use correlation, so it is not surprising that the two share the same symbol. In theory, reliability can range from 0 to 1.00, but the reliability of measures of human traits and behaviors never quite reaches 1.00.

It is the responsibility of the developer of a research instrument (e.g., employee skills test, marketing research survey, quality control process) to assess the reliability of the instrument. Those who consider using the instrument must have information about its reliability in order to make informed decisions. If those who use the instrument employ the measure to study groups and conditions that are similar to the groups and conditions used by the instrument developer, then the users of the instrument can assume that the measure has the same reliability as that reported by the measure constructor.

Understanding the Theory of Reliability

The classical theory of reliability states that an observed score, X (e.g., a score obtained on an achievement test), contains two components: a **true score** (T) and an **error score** (E). The observed score can be described as:

$$X = T + E$$

The true score reflects the *real* level of performance of what is being measured. However, this true score cannot be observed or measured directly. Based on the assumption that the error scores (E) are random and do not correlate with the true scores (T), the observed scores (X) are used to estimate the true scores. For some people (or cases), X is an *overestimate* of their T; and for other people, X is an *underestimate* of T.

Theoretically, the true score can be determined by administering the same measure over and over, recording the scores each time, and then averaging all the scores. In practice, though, individuals are tested with the same measure only once or twice at the most. Using the variance of the error scores and the variance of the observed scores, we can compute the reliability of the instrument using this formula.

$$\text{Reliability} = 1 - \frac{S_e^2}{S_x^2}$$

Where S_e^2 =   Error Variance

S_x^2 =   Variance of the observed scores

As the formula shows, a *decrease* in the ratio of the two variances causes the reliability to *increase*. This ratio can be decreased either by *decreasing* the numerator or by *increasing* the denominator. The error component is related to the way the instrument is created and to the way it is administered. Therefore, there are several ways to reduce the variance of the error scores (the numerator), such as writing good test or survey items, including clear instructions, and creating a proper environment for the testing to take place. The variance of the observed scores (the denominator) can be increased by using heterogeneous groups of examinees or by creating instruments with more items.

Methods of Assessing Reliability

This chapter describes several methods to assess the reliability of instruments: *test-retest*, *alternate forms*, and *internal consistency* methods. *Inter-rater* reliability is also discussed.

Test-Retest Reliability

The **test-retest** reliability is assessed by administering the same instrument *twice* to the same group of people. The scores from both times the instrument was administered are correlated and the correlation coefficient is used as the reliability index. The time interval between the two times the instrument is administered is important and should be reported along with the reliability coefficient. When the interval between the sessions is short, the reliability is likely to be higher than in cases when the interval between the sessions is long.

The test-retest method of assessing reliability seems the most obvious approach, because reliability is related to consistency over time. However, there are several problems involved in this method of assessing reliability. First, people need to be tested twice, which may be time-consuming and expensive. Second, when dealing with human subjects, some memory or experience from the first test is likely to affect individuals' performance on the retest. Increasing the time interval between the two testing sessions may reduce this effect, but with a longer interval, new experiences and learning may occur and affect people's performance. Generally, it is recommended that the interval between retests not exceed 6 months.

Due to the problems associated with the test-retest method, this method is not considered a conclusive measure of reliability in business and psychology. It may be used, though, in combination with other methods designed to assess test reliability.

Alternate Forms Reliability

The **alternate forms** reliability is obtained when a group of subjects is administered the two forms and the two sets of scores (from the two test forms) are correlated with each other. The alternate form method of assessing reliability is based on the assumption that if examinees are being tested twice, with two alternate forms of the same instrument, the scores on the two forms will be the same. As with test-retest, the correlation coefficient serves as the index of reliability. The two forms of the instrument should be equivalent in terms of their statistical properties (e.g., equal means, variances, and item intercorrelation), as well as the content coverage and the types of items used.

There are two major problems involved in using this reliability assessment method. The first is that the people have to be tested twice, as was the problem with the test-retest method. The second problem is that it is very difficult, and often impractical, to develop an alternate form. If the purpose is merely to assess the reliability of a single test, then the alternate form method is unlikely to be used, because it requires having a second form of the test. However, many commercial testing companies, especially those that develop personality and skills tests, construct alternate forms for other purposes. Thus, these forms can also be used to assess the test reliability. Alternate forms are useful for security reasons (e.g., every other person gets the same form to reduce copying and cheating). They are also useful in some research studies, when one form is administered as a pretest and the other form as a posttest, in order to eliminate the possible effect that previous exposure to the instrument may have on subsequent testing scores.

Measures of Internal Consistency

One major disadvantage of the two aforementioned reliability assessment methods is that the examinees have to complete the instrument twice. The **internal consistency** approaches allow the use of scores from a single administration of the instrument to estimate the reliability. In essence, each item on an instrument can be viewed as a single measurement, and the instrument can be viewed as a series of repeated measures. The internal consistency methods are based on the assumption that when an instrument measures a single basic concept, items correlate with each other and people who answer one item in a certain way are likely to answer similar items in the same way. The reliability estimates obtained by internal consistency methods are usually similar to those obtained by correlating two alternate forms. There are several methods that can be used to estimate the instrument's internal consistency.

The Split-Half Method

In this procedure, the instrument is split into two halves and the scores of the examinees on one half are correlated with their scores on the other half. Each half is considered an alternate form of the instrument. The most common way to split an instrument is to divide it into odd and even items, although other procedures that create two similar halves are also acceptable. However, dividing the instrument into the first half and the second half may create two halves that are not comparable. These two halves may differ in terms of content coverage, item difficulty, and examinees' level of fatigue and practice. To use the split-half approach, items on the instrument should be scored dichotomously, where responses that indicate a certain outcome (e.g., correct answers on a skills test) are assigned 1 point and answers that indicate the opposite outcome (e.g., incorrect answer) are assigned zero points.

The first step in the computation of the split-half reliability procedure is to obtain the scores from the two halves for each subject. The scores from one half are then correlated with the scores from the other half. Unlike the first two methods discussed (test-retest and alternate forms), this correlation is not an accurate assessment of the reliability. In fact, it underestimates the reliability because it is computed for an instrument half as long as the actual instrument for which we wish to obtain the reliability. Research has demonstrated that all things being equal, a longer instrument is more reliable. That is, if we have two instruments with similar items, but one is shorter than the other, we can predict that the longer instrument is more reliable than the shorter instrument.

In order to calculate the reliability for a full-length instrument, the Spearman-Brown prophecy formula is used. This formula uses the reliability obtained for the half-length instrument to estimate the reliability of a full-length instrument. The **Spearman-Brown prophecy formula** is:

$$r_{full} = \frac{(2)(r_{half})}{1 + r_{half}}$$

Where r_{full} =   Reliability for the whole instrument
        r_{half} =   Reliability for the half instrument (i.e., the correlation of the two halves)

Suppose we want to estimate the reliability of a 30-item instrument and the correlation of the odd-item half with the even-item half is .50. This correlation estimates the reliability for a 15-item instrument, whereas our instrument has 30 items. In order to estimate the reliability of the full-length instrument, the Spearman-Brown formula is applied as follows:

$$r_{full} = \frac{(2)(.50)}{1 + .50} = \frac{1.00}{1.50} = .67$$

In this example, the instrument developer should report the instrument split-half reliability as .67.

Kuder-Richardson Methods

G. F. Kuder and M. W. Richardson developed a series of formulas in an article published in 1937. Two of these, *KR-20 and KR-21*, are used today to measure agreement, or intercorrelation, among research instrument items. As with the split-half method, these procedures can only be used for items that are scored dichotomously (right or wrong). **KR-20** can be thought of as the average of all possible split-half coefficients obtained for a group of examinees. **KR-21** is easier to compute, but it is appropriate only when the level of difficulty of all items is similar, a requirement which is not easily satisfied.

Coefficient Alpha

This procedure was developed by Lee Cronbach in 1951. It yields results similar to KR-20 when used with dichotomous items. However, **coefficient alpha** can be used for instruments with various item formats. For example, it can be applied to instruments that use a Likert scale, where each item may be scored on a scale of 1-5. Coefficient alpha is considered by researchers to provide good reliability estimates in most situations. Readers of business research are likely to see the coefficient alpha being reported as an index of reliability because it is a popular choice among researchers.

Inter-Rater Reliability

Inter-rater reliability refers to the degree of consistency and agreement between scores assigned by two or more raters or observers who judge or grade the same performance or behavior. For example, the process of analyzing vendors' proposals or observing and rating behaviors calls for subjective decisions on the part of those who have to read the proposals or rate the behaviors. The company may assess proposals by using rubrics that include criteria such as benefits, costs, time, and past experiences with the potential vendor. To assess the reliability of the proposal selection process and the criteria used for assessing the vendors, the proposals are first read by two or more people who assign a score on each criterion using a rating scale. The scores assigned by the scorers on the different criteria are then evaluated to see if they are consistent.

The scores from two or more proposal readers can be used in two ways: (a) to compute a correlation coefficient or (b) to compute the percentage of agreement. The correlation coefficient and the percentage of agreement indicate the reliability and the consistency of the measure. A high correlation coefficient shows consistency between the readers. By providing clear guidelines for scoring, as well as providing good training, it is possible to increase the inter-rater reliability and agreement. Similarly, when two or more observers rate certain behaviors using a rating scale, their ratings are used to assess the reliability of the observation tool.

The Standard Error of Measurement

The reliability and accuracy of an instrument can be expressed in terms of the **standard error of measurement (SEM)**. The standard error of measurement provides information about the variability of a person's scores obtained upon repeated administrations of an instrument. The standard error of measurement is especially suitable for the interpretation of individual scores. Since measures of human traits and behaviors contain an error component, any score obtained by such a measure is not a completely accurate representation of the person's true performance. Instruments that are more reliable contain a smaller error component than instruments that are less reliable. The standard error of measurement allows us to estimate the range of scores wherein the true score lies. The reliability and the standard deviation (SD) of the instrument are used to compute SEM. The computation formula is:

$$SEM = SD\sqrt{1 - reliabiliy}$$

To illustrate how the SEM is computed, let's look at a numerical example. Suppose a marketing research questionnaire has a standard deviation (*SD*) of 10, and a reliability of .91. The instrument's SEM is computed as:

$$SEM = 10\sqrt{1 - .91} = 10\sqrt{0.09} = (10)(0.3) = 3$$

Relating SEM to the normal curve model, we can state that 68% of the time the examinees' *true* scores would lie within ±1SEM of their *observed* scores and 95% of the time these examinees' true scores would lie within ±2SEM of their observed scores. (See Chapter 6 for a discussion of the normal curve.) For example, when a person obtains a score of 80 on this marketing questionnaire, 68% of the time this person's true score would be expected to lie up to 3 points above or below the observed score of 80, or between 77 and 83. We can also predict that 95% of the time the person's true score would lie between 74 and 86, a range that is within 6 points of the obtained score (i.e., within ±2SEM). Clearly, it is desirable to have a small SEM, because then the band of estimate (the range within which the true score lies) is narrower and the true score is closer to the observed score.

If you inspect the formula for the computation of SEM, you would realize that the reliability of an instrument affects its SEM. A lower reliability results in a higher SEM and a wider, less precise band of estimate. Assume that the reliability of the questionnaire in our example had been .64 instead of .91. The SEM would then be computed as:

$$SEM = 10\sqrt{1 - .64} = 10\sqrt{0.36} = (10)(0.6) = 6$$

When the SEM is 6, it means that 68% of the time the examinee's true score would have been up to 6 points above or below the student's observed score of 80 (i.e., between 74 and 86), and 95% percent of the time, the true score would have been between 68 and 92. It is as if we are saying that although the person obtained a score of 80, we are 68% sure that the true score is somewhere between 6 points above to 6 points below that score.

Factors Affecting Reliability

Heterogeneity of the Group

When the group used to derive the reliability estimate is *heterogeneous* with regard to the characteristic being measured (e.g., typing speed, achievement level, or attitudes toward a new tax law), the variability of the scores is higher and, consequently, the reliability is expected to be higher. Manuals that report the instrument's reliability are likely to include information on the groups used to assess the reliability. Suppose, for example, that the group used to generate the instrument's reliability levels included employees ranging from the CEO to the newly hired mail room clerk. If the instrument is to be used with middle level managers only, the reliability of the test for the middle level managers is probably lower than that reported in the manual.

Instrument Length

As was mentioned before, all things being equal, a longer instrument is more reliable. For example, in a shorter test, the probability of guessing the right answers is higher than in a long test. Therefore, creating a longer instrument can help provide a more stable estimate of the person's responses. The split-half reliability, which uses Spearman-Brown formula, demonstrates the effect of the instrument length on reliability. It shows that a full-length instrument is more reliable than an instrument half as long. If you check the manual of a commercial test or marketing research survey, you will see that reliability levels of subsections of the instrument are usually lower than the reliability level of the whole instrument. In determining the desired length of any given instrument, though, it may be necessary to consider other variables, such as time constraints, or the ages of the prospective test takers.

Difficulty of Items

When determining the reliability of tests with right and wrong answers, tests that are too easy or too difficult tend to have lower reliability because they produce little variability among the scores obtained by the examinees. Tests where most of the items have an average level of difficulty tend to have higher reliability than tests where the majority of the items are very hard or very easy.

Quality of Items

Improving the quality of items increases an instrument's reliability. The process starts by writing clear, unambiguous items, providing good instructions for those administering and taking the instrument, and standardizing the administration and scoring procedures. Ideally, the instrument can then be field-tested with a group similar to the one intended to complete the instrument in the future. An item analysis can be performed to reveal weaknesses in the items and to help improve the instrument by reducing the error variance.

How High Should the Reliability Be?

Usually, self-made instruments or instruments created by researchers tend to have lower reliability levels than tests prepared by commercial companies or by professional instrument writers. Researchers or university professors that construct their own data collection tools may not have the time or the expertise to create the instruments, and they may not perform an item analysis or revise the items where needed.

Another point to keep in mind is that instruments that measure the *affective* domain tend to have lower reliability levels than instruments that measure the *cognitive* domain. The main reason for this phenomenon is that the affective domain behavior is less consistent than the cognitive domain behavior.

Decisions about the acceptable level of reliability depend to a great extent on the intended use of the instrument results. In exploratory research, even a modest reliability of .50 to .60 is acceptable (although a higher reliability is always preferable). For group decisions, reliability levels in the .60s may be acceptable. For example, in experimental studies that involve a comparison of experimental and control groups, individual are not usually compared; rather, group information (e.g., mean scores) is likely to be used for comparing the groups. On the other hand, when important decisions are made based on the results of the instrument, the reliability coefficients should be very high. Most commercial tests used for decisions regarding individuals have reliability levels of at least .90. Even though many managers do not have the time or the expertise to assess the reliability of the instruments they use, they should be aware of the issue of reliability in business research.

Summary

1. **Reliability** refers to the consistency of a measurement obtained for the same persons upon repeated administering of a research instrument. A reliable measure yields the same or similar results every time it is used.

2. The affective and cognitive domains are more difficult to measure reliably than are physical traits.

3. The *real* level of performance for any individual, or the **true score** (*T*), cannot be observed directly. The **observed score** (*X*) is likely to *overestimate* or *underestimate* the true score for any given individual. The observed score equals the sum of the true score and the **error score (E)**.

4. The reliability of a measure can be represented by the formula:

$$\text{Reliability} = 1 - \frac{s_e^2}{s_x^2}$$

5. Methods for *decreasing* the error component include writing good items, giving clear instructions, and providing an optimal environment for the test takers. Methods of *increasing* the variance of the observed scores include using heterogeneous groups of examinees and writing longer tests.

6. The reliability of a particular measure may be assessed using these methods: *test-retest, alternate forms*, and *internal consistency* approaches.

7. **Test-retest** reliability is assessed by administering the same instrument *twice* to the same group of people. The scores of the people from the two sessions are correlated and the correlation coefficient is used as the reliability index.

8. The **alternate forms** reliability is obtained when a group of examinees is administered two alternative forms of the instrument and their two scores are correlated with each other. The correlation between the two alternate forms serves as the index of reliability.

9. Measures of **internal consistency** use the scores from a single administration of the instrument to estimate its reliability. In this method, each individual item becomes a single measurement while the instrument as a whole is viewed as a series of repeated measures. Internal consistency methods include the *split-half, Kuder-Richardson* methods (*KR-20* and *KR-21*), and *coefficient alpha*.

11. In the **split-half** method, the instrument is split into two halves and the scores of the examinees on one half are correlated with their scores on the other half. Each half is considered an alternate form of the instrument.

12. In order to calculate the reliability for a full-length instrument, the **Spearman-Brown prophecy formula** is used with the split-half reliability approach. This formula uses the reliability obtained for the half-length instrument to estimate the reliability of a full-length instrument. The Spearman-Brown prophecy formula is:

$$r_{full} = \frac{(2)(r_{half})}{1 + r_{half}}$$

13. The reliability of an instrument can be assessed using the **KR-20** and **KR-21 formulas** that are used to measure agreement, or intercorrelation, among test items. Scores obtained from a group of people who have taken the test one time can be used to obtain this reliability estimate.

14. **Coefficient alpha** can be used to assess the reliability of instruments with different types of item formats using scores obtained from a single testing of the instrument.

15. **Inter-rater** reliability refers to the degree of consistency and agreement between scores obtained by two or more raters or observers who judge or assess the same performance or behavior.

16. The **standard error of measurement (SEM)** measures the reliability and accuracy of the instrument in relation to its ability to accurately estimate the range of scores within which the true score lies. SEM is calculated using this formula:

$$SEM = SD\sqrt{1 - reliabiliy}$$

17. A smaller SEM allows for a more accurate estimate of the true score and, therefore, provides a more reliable measure. Instruments with higher levels of reliability have lower standard errors of measurement than less reliable tests.

18. Using the normal curve, we can state that 68% of the time the examinees' *true* scores would lie within $\pm 1 SEM$ of their *observed* scores and 95% of the time the examinees' true scores would lie within $\pm 2 SEM$ of their observed scores.

19. Factors such as the *heterogeneity* of the group, the test *length*, and the *difficulty* and *quality* of the items affect the reliability of the measure.

20. Self-made instruments and those created by researchers to be used in a research project tend to have lower reliability levels than commercial instruments or tests that were created by experts.

21. Instruments that measure the *affective* domain tend to have lower reliability levels than those that measure the *cognitive* domain because the affective domain behavior is less consistent than the cognitive domain behavior.

22. Important decisions should not be made using a score from a single instrument because each instrument contains a certain level of error. Instead, scores from multiple measures should be used.

23. Decisions about the acceptable level of reliability depend largely on the intended use of the instrument results.

Chapter 14

Validity

The **validity** of a test refers to the degree to which an instrument measures what it is supposed to measure and the appropriateness of specific inferences and interpretations made using the test scores. It is not sufficient to say that a test is "valid"; rather, the *intended use* of the test should be indicated. Keep in mind that validity is not inherent in the instrument itself and that an instrument is considered valid for a particular purpose only. For example, a test that is a valid a measure of quality for *manufactured* products may not be a valid measure of quality in the *service* industry. Validation of a test involves conducting empirical studies where data are collected to establish the instrument's validity. A valid test is assumed to be reliable and consistent; however, a reliable test may be valid only for a specific purpose. There are four basic types of validity: *content* validity, *criterion-related* validity, *construct* validity and *face* validity.

Content Validity

Content validity describes how well an instrument measures a representative sample of behaviors and content domain about which inferences are to be made. In order to establish the content validity of the instrument, its items are examined and compared to the content of the unit to be measured, or to the behaviors and skills to be measured.

Managers who are assigned the responsibility of choosing a series of standardized, commercial pre-employment tests for their organizations need to compare the items on the tests to their criteria for hiring and job performance and make sure they match. Because hiring criteria are likely to differ from organization to organization, a particular standardized test may have a high content validity for some organizations but a low content validity for others.

Managers who create their own research instruments should make sure that items on the instrument correspond to the information they gain from the study. For example, a manager who wants to measure her employees' attitudes towards implementing a new cost cutting process may not want to ask questions about the employees' personal opinions of the CEO, if this is not relevant to the implementation plans. Irrelevant items such as this lower the validity of the test.

Criterion-Related Validity

The process of assessing the **criterion-related** validity of a measure involves collecting evidence to determine the degree to which the performance on a measuring instrument is related to the performance on some other external measure. The external measure is labeled as the *criterion*. As part of the process to assess the criterion-related validity of the instrument, test developers can correlate it with an appropriate criterion. The correlation coefficient is called the **validity coefficient** and it is used to indicate the strength of the relationship between the instrument and the criterion. There are two types of criterion-related validity: *concurrent* validity and *predictive* validity.

Concurrent Validity

Concurrent validity is concerned with the evaluation of how well the test we wish to validate correlates with another well-established instrument that measures the same thing. The well-established instrument is designated as the *criterion*. For example, a newly created short version of a well-established test may be correlated with the full-length test and a high correlation between the two tests indicates that they measure similar characteristics, skills, or traits. In order to establish concurrent validity, the two measures are administered to the same group of people, and the scores on the two measures are correlated. The correlation coefficient serves as an index of concurrent validity.

To illustrate, suppose a researcher develops a new IQ test that takes 30 minutes to administer and 30 minutes to score. This is much faster than the commonly used IQ tests. In order to establish the concurrent validity of the new IQ test, the researcher may correlate it with a well-established IQ test by administering both tests to the same group of people. High and positive correlation of the new test with the established IQ test would lend support to the validity of the new test.

Predictive Validity

Predictive validity describes how well an instrument predicts some future performance. This type of validity is especially useful for pre-employment tests that are designed to predict some future performance. The test to be validated is the *predictor* (e.g., a pre-employment skills test) and the future performance is the *criterion* (e.g., performance of the employee). Data are collected for the same group of people on both the predictor and the criterion and the scores on the two measures are correlated to obtain the validity coefficient. Unlike concurrent validity where both instruments are administered at about the same time, predictive validity involves administering the predictor first, while the criterion is administered later in the future.

Suppose a market researcher wants to establish the predictive validity of a purchase intent survey for a new brand of coffee that a company is about to introduce into the market. The purchase intent survey is administered to a sample of 400 consumers. A month later, the same 400 consumers are asked whether or not they purchased the new coffee. Next, the purchase intent scores are correlated with the consumers' responses to whether or not they actually bought the coffee. A high positive correlation indicates that the purchase intent survey has a high predictive validity, because it predicted accurately the consumers' likelihood of purchasing the new coffee product.

You should keep in mind that tests that are intended to predict future performance may provide incomplete information about the criterion. For example, the purchase intent survey may not always predict whether or not a consumer still purchases the coffee later. The reason is that this test may measure intent at the time, but cannot account for future factors such as changes in competition, consumers' changes in preferences, changes in consumers' income, or a wide variety of other variables that could cause consumers to change their minds.

Construct Validity

The term *construct* is used to describe characteristics that cannot be measured directly, such as intelligence, attitude, and trust. **Construct** validity is the extent to which a test measures and provides accurate information about a theoretical trait or characteristic. The process of establishing the instrument's construct validity can be quite complicated. The researcher administers the test to be validated to a group of people, and then collects other pieces of data for these same people.

For example, let's say that a new scale has been developed in order to measure trust in an organization. To demonstrate that the scale indeed measures organizational trust, the researcher first administers the scale to a group of people, and then collects additional information about them. Those who score low on the organizational trust measure are considered to have a low level of trust in those with whom they work and are expected to exhibit behaviors and responses that are consistent with low levels of organizational trust. Conversely, those who score high on the test are expected to behave in ways that are compatible with a high level of organizational trust. Thus, establishing construct validity consists of accumulating supporting evidence. Evidence for construct validity is not gathered just once for one sample; rather, it is collected with the use of many samples and multiple sources of data.

Face Validity

Face validity refers to the extent to which an instrument *appears* to measure what it is intended to measure. The extent to which a test appears valid to the examinees and to other people involved in the research process may determine how well the instrument is accepted and used. Additionally, face validity helps to keep people motivated and interested, because they can see the relevancy of the test to the perceived task. For example, a test with a high face validity that is used to screen a pool of applicants for certain positions is quite defensible as an appropriate instrument because applicants can see the test as relevant and perceive it as an appropriate measure.

Face validity is likely to be assessed based on a superficial inspection of an instrument. However, this inspection is not sufficient. The mere *appearance* of face validity is not a guarantee that an instrument is *valid* and that it truly measures what it is supposed to measure. You should be aware of the fact that face validity is not always found in discussions of validity, and it may not be considered by all to be as important as the other types of validity.

Assessing Validity

Although we have identified several different types of validity, they are not necessarily separate or independent of each other. Establishing the measure's validity usually involves a series of steps of gathering data. Information provided by the developer of the instrument about its validity should include a description of the sample used in the validation process. Ideally, the characteristics of this sample are similar to those of future test takers.

Assessing the content validity of an instrument does not involve numerical calculation. Rather, it is a process of examining the instrument in relation to the content it is supposed to measure. In measuring criterion-related validity, the validity coefficient is used to describe the correlation between an instrument and a criterion. To be useful, the criterion has to be reliable and appropriate. The process of establishing the construct validity of an instrument includes the use of statistical methods (e.g., correlation), as well as procedures for gathering and comparing various measures.

Test Bias

Standardized tests, especially those used for employment screening, admission, certification, and placement are viewed at times as being *biased* against one group or another. A test is considered biased if it consistently and unfairly discriminates against a particular group of people who take the test. For example, certain tests are said to be gender biased, usually discriminating against female examinees. Other tests may be considered biased against racial or cultural minorities.

Norm-referenced tests are constructed to discriminate *among* examinees of different ability levels. This type of discrimination is not to be confused with the notion of test bias, where a test systematically discriminates against a particular group of examinees.

Summary

1. **Validity** refers to the degree to which an instrument measures what it is supposed to measure and the appropriateness of specific inferences and interpretations made using the instrument scores. The intended use of the instrument should be indicated, because an instrument is considered valid for a particular purpose only.

2. The three basic types of validity are: *content* validity, *criterion-related* validity, and *construct* validity.

3. **Content validity** refers to the adequacy with which an instrument measures a representative sample of observations and content domain about which inferences are to be made. In order to establish the content validity of the instrument, its items are examined and compared to the content of the unit to be tested, or to the behaviors and skills to be measured.

4. Instruments have **criterion-related validity** with respect to the relationship of scores on two separate measures. One measure is the newly developed instrument, and the other measure serves as a criterion. There are two types of criterion-related validity: *concurrent* and *predictive*.

5. The correlation coefficient between the instrument and the criterion is called the **validity coefficient** and it indicates the strength of the relationship between the two measures.

6. Assessing the instrument's **concurrent validity** involves evaluating the degree to which the results from the instrument correlate with another well-established instrument that measures the same thing.

7. An instrument has **predictive validity** if it can successfully predict future performance in a given area. The newly-developed instrument is called the *predictor* and the future performance is the *criterion* that is used to establish the predictive validity of the predictor.

8. **Construct validity** refers to the extent to which an instrument measures and provides information about a theoretical trait or characteristic. To establish the construct validity of an instrument, it is necessary to collect additional data over a period of time and to correlate these data with the research results.

9. **Face validity**, which is not always recognized as a formal type of validity, refers to the extent to which an instrument *appears* to measure what it is intended to measure.

10. A test is considered *biased* if it systematically discriminates against a particular group of examinees.

Part Seven

Conducting Your Own Research

Chapter 15

Planning and Conducting Research Studies

Before researchers carry out studies, they need to plan and map out their steps. After the study is conducted, most researchers write a report that summarizes the study. This chapter focuses on the process of writing research proposals for quantitative studies and on reports that describe these studies.

Both qualitative and quantitative studies require a clearly articulated research proposal prior to beginning the study. However, different research paradigms may follow different guidelines and require different approaches to the process of planning, conducting, and reporting research studies. While quantitative studies demand more detailed plans, proposals for conducting qualitative studies may be less structured. Since this textbook is about quantitative statistics, the discussion here focuses on proposals and reports of quantitative studies. In this chapter, we focus on guidelines for students and researchers who conduct studies that examine quantitative data. Keep in mind, though, that your university or college probably has its own specific set of rules and guidelines that may differ from those described in this chapter.

Students who are writing proposals to meet degree program requirements, such as theses or dissertations, would probably need to follow specific guidelines given to them by their committees. Often businesses and other organizations that do a large number of research projects are likely to have their own guidelines for creating proposals. Therefore, the discussion here is geared mainly towards students who plan to conduct research projects as a class assignment, or for practitioners who would like to study their own settings.

After a study is completed, it is described in a research report. Although the research proposal and the research report share common elements, they differ in several ways. For example, proposals include only three main chapters: *Introduction*, *Literature Review*, and *Methodology*. By comparison, research reports include these chapters *plus* two additional chapters: *Results* and *Discussion*. Reports may also include an *Abstract* that summarizes the report and appears at the beginning. Another key difference between proposals and reports is the tense used. Proposals are written using *future* tense whereas reports use *past* tense. Both proposals and reports also include a chapter called *References*, which lists all the references cited or quoted in the text. Additionally, an *Appendix* may be found in both proposals and reports.

When writing a proposal or report, you may be asked to use a particular writing style. The most well known writing style and the one used by most universities and many businesses is the one described in the *APA Publication Manual*.[1] However, other styles may be used as well and you should check to see which style you should follow.[2]

---

1. **A HINT**: At the time this book is written, the most recent edition of the *APA Publication Manual* (published by the American Psychological Association) is the fifth edition, dated 2001. As you read this book, check to see if there is a newer edition. The most recent APA publication guidelines may also by found on the Internet at http://www.apastyle.org.
2. **A HINT**: Another writing style that is used by university students and faculty is called The Chicago Style, which was originally written by Turabian in 1937 and has been updated several times since then. The *MLA Handbook*, published by The Modern Language Association of America, is also used in some cases.

Before writing your research proposal, you should investigate your topic by reading about it as much as possible. By reviewing the literature you become well informed about your topic, gather background information, learn about current trends and theories related to your topic, and identify gaps and controversies in the literature. All of these should help you sharpen your focus and select your own specific research topic. The literature review process can also prevent you from unintentionally duplicating other studies, and will also help you avoid other researchers' mistakes, as well as benefit from their experience.

There is a wealth of information available electronically on the Internet. The electronic data search techniques change and are updated at a rapid pace; it is probably a good idea for you to consult with your librarian in order to learn about the most recent techniques for electronic literature search.

Note that in discussing the different parts of the proposal and report in this book, the term *chapter* is used to describe the main parts of the proposal and report. The word *section* is used to denote a subpart of a chapter.

Every researcher undertaking a research study should be cognizant of ethical considerations involved in research. Before discussing the research proposal, we briefly review the ethics of research. This chapter then continues with a discussion of research proposals, followed by a discussion of research reports.

Research Ethics

When planning your study, you should be concerned with **research ethics**. The rights of those you study should be protected at all times. As a researcher, you have to ensure that the participants are well informed of the nature of the study and that you have not placed them in risky situations. Adult participants should provide their consent to participate in the study, while parents or legal guardians should provide consent for minors under their care. The study's participants also should be given an opportunity to withdraw from the study after it has started. People's request for privacy should be honored and their confidentiality should be assured, when requested.

Ethical considerations are especially important in experimental studies, where participants undergo planned interventions. However, practitioners who conduct research in their own settings should also maintain high ethical standards and be aware of all possible consequences of their studies. When you conduct your own work-related research, you should ensure the rights, welfare, and well-being of the study's participants. You may want to check with your manager before undertaking your study and secure his/her permission to conduct the study.

Several professional associations provide guidelines for their members regarding studies that involve human subjects.[3] Institutions, such as universities, as well as government offices and businesses, may request that all research proposals undergo a review by a human subjects review board as part of the proposal approval process.

The Research Proposal

After deciding on a topic for your study, your next step is to write a research proposal. A **proposal** may be viewed as the blueprint for the study. It provides a rationale for the study and an explanation of the reasons the study should be conducted. A well written, carefully planned proposal helps you plan ahead, anticipate your needs, and outline a realistic timetable.

---

3. **A HINT**: See, for example, the guidelines of the American Psychological Association webpage at http://www.apa.org/ethics/. See also: The American Educational Research Association (2002). *Ethical standards of the American Educational Research Association: Cases and commentary*. Washington, DC: Author.

A typical proposal has three chapters: *Introduction, Literature Review,* and *Methodology*. A list of sources used in the proposal is also included in **References**. All proposals are expected to include these three parts, whether they are submitted to fulfill requirements for a degree or to conduct a study at work. In some cases, the *Introduction* and *Literature Review* are combined into one chapter, called *Introduction*. Regardless of the number of chapters required, all proposals should have an introductory chapter that includes a statement of purpose or research questions and any hypotheses, as well as a brief review of the literature. Additionally, a proposal should contain a description of the study's methodology and a list of the references cited. The *Methodology* chapter may also include a section entitled *Data Analysis*. Information about data analysis can be included either as a separate section in *Methodology* or as part of the *Procedure* section.

Following is a description of the main components of the research proposal: *Introduction, Literature Review*, and *Methodology*. A brief discussion of *References* is also included.

Introduction

The **Introduction** chapter introduces the study by stating the problem to be investigated, the purpose of the study, the rationale for conducting the study, and the study's potential contributions to the field. This chapter also includes the research questions and any hypotheses stated by the researchers prior to the study.[4] The background of the problem should be briefly developed in this chapter, but the main discussion of background information should be included in the *Literature Review* chapter.

The *Introduction* chapter includes a *statement of the problem* that is written as a question or a declarative statement and is usually placed at the beginning of the chapter. For example, a declarative statement may state that the study was designed to investigate the effect that a decline in interests rates has on the number of new home mortgages. Examples of problems stated as questions are: What are the advantages of providing six additional months of training to new employees? How do the differences in state government regulations for starting a business affect the success rate of new businesses?

The *Introduction* chapter should include a rationale and a clear explanation of the need for studying the problem and for finding solutions to the problem. After reading this explanation, the reader should be able to understand the potential contribution of the proposed research to business practice or theory. For example, with regard to increasing interest rates, has this study been done before? If so, have conditions changed enough to justify doing the study again? If the study has not been done before, why not? Will anybody care that the study is being done? As to providing more training, it may be argued that on-the-job experience is more valuable than more training. However, if the company has had a difficult time keeping new employees because they complain about being ill-prepared, a study attempting to measure the value of more training may be useful. The third topic dealt with the state government regulations associated with starting new businesses in different states. One may argue that a good business can be started anywhere if the new business is fulfilling an unmet need or want. Others contend that states with fewer business regulations are easier places for new ventures to succeed. Therefore, a study should be undertaken to compare the success rates of new businesses with different levels of state government regulation.

Many proposals also include research hypotheses, especially those that propose experimental studies. The exact placement of the hypotheses may differ. Most guidelines require placing them in the *Introduction* chapter, but others may suggest placing the hypotheses in the second or third chapter (*Literature Review* or *Methodology*) of the proposal. We suggest that you check the specific guidelines given to you for further directions about the placement of the study's hypotheses.

---

4. **A HINT**: See Chapter 2 for a discussion of research questions and hypotheses.

The *Introduction* chapter also includes a brief review of selected sources that are most related to the topic. Those references are discussed in greater detail in the *Literature Review* chapter that follows the *Introduction*. Proposals may also include definitions of key terms, assumptions, and limitations of the study.

Literature Review

The **Literature Review** chapter summarizes literature related to the topic being investigated. In the proposal, the review of related literature tends to be limited in scope, citing briefly a small number of studies. Later, when writing the research report, this section is expanded. Literature reviews in dissertation and thesis proposals are expected to include the most important studies on the topic. In less formal proposals, such as those written as part of class research projects, the literature review is not likely to be comprehensive due to time constraints. When the literature review in proposals is comprehensive and includes a discussion of a number of subtopics, it is recommended that a summary of the review be included at the end of this chapter.

When writing the review, it should be organized by topic, rather than as an annotated bibliography or a series of summaries of articles, reports, or books. As the writer of the proposal, it is your responsibility to synthesize the research on your topic and point the reader to controversies in the field, as well as similarities, agreements, or disagreements among researchers who have conducted research on your topic. Existing gaps in knowledge and practice should also be noted. In some cases, you may want to include a critique of the studies you review and point to their shortcoming and contributions.

The information presented in the *Literature Review* should be properly attributed to its authors to avoid plagiarism. Sources must be acknowledged whether quoted directly or paraphrased. You should summarize key ideas, findings, and conclusions of other researchers. It is best to quote very little, if at all, and quote only phrases or ideas that are so well stated you feel you cannot summarize them accurately. Try to keep the tone of your writing objective and unbiased and present a balanced discussion of all views, even those you may personally oppose.

A number of studies that are discussed in the *Literature Review* chapter would most likely be discussed again in the *Discussion* chapter of your research report. In that chapter, results from your own study should be examined and related to the existing body of knowledge in the field.

Methodology

The **Methodology** chapter in the proposal is designed to describe your plan of action and to clarify to the reader how you are going to investigate the research questions and test the hypotheses. The description of your methodology should be specific enough to communicate to the reader that you have carefully planned every step of your study. The three main sections of this chapter are: *Sample* (or *Participants*), *Instruments* (also called *Data Collection Tools*, *Tests* or *Measures)*, and *Procedure*.[5] Sections about *Design* or *Data Analysis* may also be included in this chapter.[6]

When planning the specifics of the study, you should ask yourself questions such as: Do I have the expertise, resources, and know-how to carry out the study? Is my study feasible? Have I set a realistic timetable to design, conduct, and complete the study? Can I obtain the cooperation and collaboration of others, if needed? Do I need permission to conduct the study? What data collection instruments should I use? How can I recruit participants for my study?

---

5.   **A HINT**: In some textbooks and journal articles, the term *subjects* is used in place of *sample of participants*. Note, however, that currently most guidelines recommend using the terms we use in this book, namely *Sample or Participants*, rather than *Subjects*.

6.   **A HINT**: Some writing guidelines further divide the *Methodology* chapter (especially in experimental studies) into additional sections, such as: **Materials, Independent Variables**, and **Dependent Variables**.

Sample

The **Sample** section describes the participants of your study. In most studies, your participants are likely to be people, but a sample can comprise of a group of cases or items. You should present information related to the sample, such as how the sample will be selected, the size of the sample, and relevant characteristics about the sample. You, as the researcher-author, have to decide which characteristics are relevant for your study. For example, for human subjects in a marketing study, family income, age, or social class may be considered important. If you are studying a sample of businesses, then sales, number of employees or industry affiliation may be relevant.

Obviously, the *exact* information about the sample in your own study (e.g., the mean age or the number of executives in each group) may not be known until you actually conduct the study. However, the *Sample* section should communicate your plans and intentions and provide a general description of the study's participants.

Instruments

In the **Instruments** section, the data collections tools that you plan to use in the study should be clearly described and their purposes explained. If you plan to use existing instruments that were developed by others, their reliability and validity should be reported.[7] Additional information about the instruments may also be reported when available. For example, you may describe the number and type of items used and the length of time required to complete the instrument. Check for copyright information and for permission to use the instrument or to include it in your proposal.

If you plan to develop a data collection instrument (e.g., a questionnaire or an aptitude test), explain how you plan to construct it and the type of items you will use. You should also discuss how you plan to assess the instrument's reliability and validity and whether you plan to pilot test it first before using it in your full-scale study. It is also advisable to include sample items of your proposed instrument. You can include the complete instrument in the proposal's *Appendix*.

Procedure

The **Procedure** section describes how the study will be conducted. It explains, in as much detail as possible, what will happen and how you will carry out the proposed investigation. This section is especially important in experimental studies that require a detailed description of the intervention. Examples of information to present in this section include a description of the training required to implement a new experimental method and the types of instructions to be provided to respondents who are asked to complete a survey. This section should also contain a realistic timetable for the different phases of the study.

Data Analysis

A description of your plans for organizing and analyzing the data to be collected in the study should be included in the **Data Analysis** section. You have to decide in advance which statistical tests and techniques you will use to enable you to investigate your research questions and test your hypotheses.

---

7. **A HINT:** Several publications provide information about published tests. For example, the Buros Institute publishes the *Mental Measurement Yearbook* and *Tests in Print*. (See http://www.unl.edu/buros/; test reviews online are also available in this website.) *Test Critiques* (Vol. 11, 1998), as well as other sources, provide information about existing instruments. Check with your librarian. See also Chapters 13 and 14 in this book for additional information about reliability and validity.

In descriptive studies, the data analysis may include tables to summarize basic descriptive statistics, such as percentages, means, and ranges. Graphs and charts are also likely to be used in such studies. Other studies, especially those using inferential statistics, may require statistical techniques, such as the t test and analysis of variance (see Chapters 10 and 11, respectively).

References

The last chapter in a proposal is *References*. Here you list all the references cited or quoted in the proposal. The exact way to list the references depends on the writing style used. For example, *APA Publication Manual* requires that you list all the references in alphabetical order according to the authors' last names. Other styles may specify that references be listed in the order they are cited in the text. The exact rules and guidelines for listing each individual reference also vary depending on the writing convention used. Regardless, all references cited in the text should be listed in the *References* chapter; and all references listed should have been cited in the text.

The Research Report

After conducting your research study and analyzing the data collected, you are now ready to write your report. As with the proposal, specific guidelines for writing research reports may vary, depending on the nature and purpose of the report. For example, if you have to write the report as part of a class research project, your instructor may give you particular guidelines to follow. If this is a thesis or a dissertation study, your committee will ask you to follow your university's guidelines. In general, though, all quantitative research reports are likely to have similar components.

Research reports usually have six chapters: *Introduction, Literature Review, Methodology, Results*, and *Discussion*. The sixth chapter, *References*, includes a list of references cited or quoted in the first five chapters. And, as was the case with the proposal, the references listed in that chapter should correspond to those cited in the text.

The first three chapters of a research report are the same as those found in a proposal (i.e., *Introduction, Literature Review*, and *Methodology*), but they are longer, more detailed, and better developed. In a report, the *Literature Review* chapter usually includes more references and citations than in a proposal. This is true especially in theses and dissertations, where the author is expected to include *all* references related to the research topic. When writing the *Methodology* chapter in the report, you are likely to include more specific information about the study. For example, you can now report the exact number of those who participated in the study and how many responded to a questionnaire you administered. And, unlike the stage of writing the proposal, now you have the results of the study and can write the *Results* and *Discussion* chapters.

At times, the *Introduction* and *Literature Review* chapters are combined into one chapter, titled *Introduction*.[8] For example, some journal editors advise authors who are interested in submitting manuscripts to their journals to combine the two chapters. However, dissertations and theses guidelines typically advise students to keep the two chapters separate.

---

8.   **A HINT**: According to APA publication manual, there is no need to type the word *Introduction* because its placement at the beginning of the manuscript identifies it as the introduction. However, you should check the specific typing guidelines given to you.

You may also be asked to write an *Abstract*, which is usually found at the beginning of the report, before the introduction. The **Abstract** summarizes the study and focuses on the study's research problem, methodology, main results, and major conclusions. *Abstracts* are usually limited in length (i.e., number of words), from about one short paragraph of 100 words, to about a page-and-a-half (or 1000 words). Because the length of the *Abstract* is limited, it has to be succinct and present only the most important points. If you plan to submit an article for publication, check the guidelines specified by the journal editors.

Many reports, especially theses and dissertations, include an *Appendix* at the end of the report. The *Appendix* includes information that is too lengthy or too specific to be included in the text of the report. For example, the *Appendix* may include the complete survey used in a study or a letter asking a manager for permission to interview her employees.

Results

The **Results** chapter presents the study's findings. This chapter includes numbers, tables, and figures (i.e., charts and graphs). The information presented and conveyed to the reader in this chapter should be written objectively, factually, and without expressing personal opinion. For example, you should not make statements such as, "We were disappointed to see that interest rates did not affect the number of new mortgages sold, since we hoped that would be the case."

A good way to organize and discuss your findings in this chapter is to reiterate the hypotheses, one by one, and present the data that were collected to test each hypothesis. It is your decision as to what data to present in a narrative form and what to present in tables or figures. Quite often, the tables and figures are accompanied by a narrative explanation. There is no need to describe in words *everything* that is presented in a numerical or visual form. Instead, it is the responsibility of the writer of the report to "walk" the reader through the numerical and visual information. As the author, you should highlight the main findings, point to trends and patterns, and guide the reader through the information you present. For example, in a table displaying results from four independent-samples t tests, you can state that the second t value, which was used to test the second research hypothesis, was statistically significant at $p<.01$, and that the mean of the experimental group was 8 points higher than the mean of the control group. There is no need to repeat in the narrative all the numerical information reported in the tables. Or, suppose your *Results* chapter includes a double-bar graph that is used to show trends and differences in the percentages of male and female Chief Financial Officers (CFOs) in public corporations, private corporations, and nonprofit organizations. You may explain that the trend is an increase in the percentage of male CFOs in non-profits, whereas the percentage of female CFOs is decreasing in private corporations.

As to the actual typing of tables and figures, consult the guidelines given to you. Each style has different requirements and those requirements can be quite specific. For example, according to APA style, the title of a table should be typed *above* the table while the caption (i.e., title) of a figure should be typed *below* the figure.

The tables and charts you construct should be easy to read and understand. In all likelihood, the computer printouts produced by the statistical program you use are not going to be "reader friendly" and you will probably need to retype them following the guidelines given to you.

Discussion

Results from the study are discussed, explained, and interpreted in the **Discussion** chapter. The results are examined to determine whether the study's hypotheses were confirmed. This chapter allows you to offer your interpretation and explain the meaning of your results. If the findings are different from those that were predicted by the hypotheses, you have to provide tentative explanations for those discrepancies. For example, some common explanations for unexpected results in a study are that the sample size was too small, the study was too short, directions given to participants were not followed properly, the instruments were not valid or reliable, or the survey response rate was too low. In some studies, one may speculate that the responses given by the participants were contrary to what was expected because people were dishonest in their responses or were reluctant to share certain sensitive information with others.

Often, the study's shortcomings are discussed in a section called **Limitations of the Study**. For example, you may explain that the results of the study should be generalized only to other groups with characteristics similar to those of the study's participants. Or, you may state as a limitation the fact that people may not have been honest in their responses.

Besides discussing the results from your own study, you should include in this chapter a discussion of your findings in relation to findings from other researchers. You should also point to examples where your own research supports or contradicts other researchers whose work was discussed in the *Introduction* and *Literature Review* chapters. By doing so, you demonstrate how your study relates to the field and to its knowledge base.

Other sections, such as **Conclusions, Recommendations, Implications** (or **Implications for Practitioners**), and **Suggestions for Further** (or **Future**) **Research** may follow the discussion of the findings and be included in the *Discussion* chapter. Again, consult the guidelines given to you to find out what you are expected to include in the research report.

Summary

1. Different research paradigms follow different guidelines and require different approaches to the process of planning, conducting, and reporting research studies.

2. *Quantitative* studies demand a more detailed research plan, compared to proposals for *qualitative* studies. This chapter discusses how to plan, conduct, and report *quantitative* research, with a focus on numerical data.

3. When writing a proposal or report, researchers usually follow a specific writing style, such as *APA* style.

4. Researchers, including those studying their own practice, should follow *ethical* principles. This is especially important in experimental studies where participants undergo a planned intervention. The rights of the study's participants should be protected at all times. Other guidelines are also discussed in the chapter.

5. The research **proposal** can be viewed as the blueprint for the study. It also provides a rationale for the study and an explanation for the reasons the study should be conducted.

6. A research proposal includes the following chapters: *Introduction, Literature Review, Methodology*, and *References*.

7. Reading as much as possible about your topic will assist you in narrowing down and selecting your specific topic and in writing the literature review. It will also prevent you from unintentionally duplicating other studies and help you select methods and procedures for your study.

8. The **Introduction** chapter of your proposal should provide a brief background of the problem and explain the significance of the topic and its potential contributions to the profession. This chapter should also include a rationale for your study in order to convince the reader that your topic is worthwhile.

9. Your proposal should include a *statement of the problem* that explains the question to be explored. It should be in a form of a declarative statement or a question.

10. Most proposals include hypotheses. This is true especially in proposals that are written to propose an experimental study.

11. *Definitions*, *assumptions*, and *limitations* may also be included in a proposal.

12. The **Literature Review** chapter summarizes research related to the topic being investigated. All information presented in the review should be properly acknowledged and attributed to its authors to avoid plagiarism.

13. The **Methodology** chapter is designed to describe your plan of action and to clarify to the reader how you are going to answer the research questions and test the hypotheses.

14. Information about those who will participate in the study and their demographic characteristics is found in the *Methodology* chapter, under **Sample** (or **Participants**).

15. A clear description of the instruments to be used in the study should be included in the *Methodology* chapter under **Instruments** (also called **Data Collection Tools**, **Tests** or **Measures**). When appropriate, sample items, as well as information about instruments' reliability and validity, should be included.

16. The *Methodology* chapter also includes a **Procedure** section that describes how the study will be conducted. This section is especially important to include in experimental studies that require a detailed description of the intervention.

17. The **Data Analysis** section should describe how you plan to organize and analyze the data to be collected in the study.

18. The last chapter in a proposal is **References**. All references cited or quoted in the text should be listed in this chapter.

19. Proposals and research report may also include an *Appendix* after *References*. The *Appendix* includes information that is too lengthy or too specific to be included in the text.

20. Most research reports include the following chapters: *Introduction*, *Literature Review*, *Methodology*, *Results*, *Discussion*, and *References*.

21. The first three chapters of the report are the same as those in the proposal (i.e., *Introduction*, *Literature Review*, and *Methodology*). However, these chapters in the report are longer, more detailed, and better developed than in the proposal.

22. An **Abstract**, summarizing the study, may also be included in a research report. It is usually placed right at the beginning of the report (before the *Introduction*).

23. The information in the **Results** chapter should be reported objectively, factually, and without expressing personal opinion. This chapter tends to be comprised of words, numbers, tables, charts, and figures. A good way to organize your findings is to reiterate the hypotheses (or research questions) one by one and present the data that were collected to test each hypothesis.

24. The results from the study are discussed, explained, and interpreted in the **Discussion** chapter. This chapter refers back to the study's research questions and hypotheses and discusses them. It also places the results from the study in relation to findings from previous studies.

25. The research report may also include **Conclusions**, **Recommendations**, **Implications**, or **Suggestions for Further Research**.

Chapter 16

Choosing the Right Statistical Tests

After researchers collect their data, they have to analyze their data in order to answer the study's questions and test its hypotheses. This chapter provides an opportunity for you to practice an important skill; that of choosing the proper statistical test to analyze the data you collected.

The chapter includes a decision flowchart that displays the various statistical tests covered in this book. The first level of the flowchart lists measurement scales of data. There are two choices: (a) nominal, and (b) interval/ratio scales. The second level of the flowchart displays the types of hypotheses to be tested. There are two types of hypotheses: (a) hypotheses that measure differences between groups or sets of scores, and (b) hypotheses that measure association between variables.

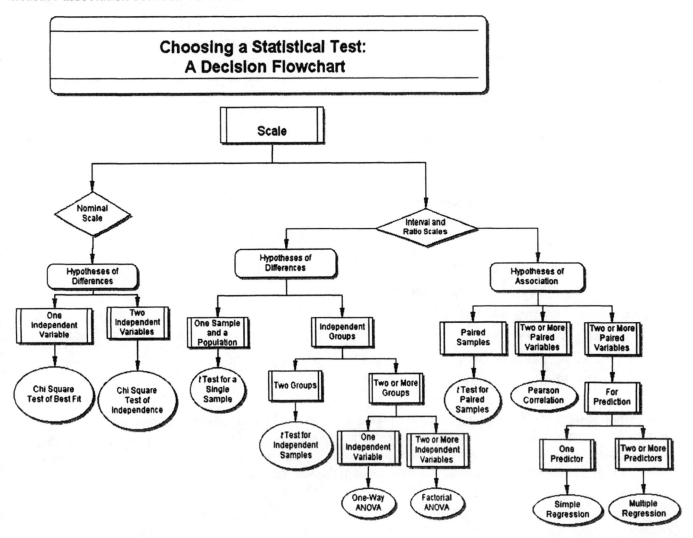

In general, statistical tests may be classified into those designed to test hypotheses of association and those designed to test hypotheses of difference. It may be easier for you to distinguish between these two types of tests if you remember the following: tests that are designed to measure association between variables can indicate the presence or absence of an association and indicate the degree (or extent) of such an association. For example, the Pearson correlation that is used to test hypotheses of association can also provide information about the degree of association between two paired variables. This is done through the use of the correlation coefficient r. Tests that are designed to measure differences can also indicate the presence of relationship between independent and dependent variables, but these tests cannot indicate the degree of the relationship. For example, a t test for independent samples may be used to measure the relationship between, let's say, gender and attitude toward a brand name, but it cannot *quantify* the magnitude of these relationships.

The third level of the flowchart asks you to decide whether the groups or variables in the study are *independent* or *paired*, and whether there are *one* or *more groups* or *variables*. The final level in the flowchart includes a series of circles that list the various statistical tests that are introduced in the book.

To help you learn how to select the right statistical test to analyze your data, following are 14 research scenarios that provide you with practice opportunities. After reading each scenario, decide which statistical test should be used to analyze the data and answer the research question that is stated or implied in that scenario. You can check your answers with those provided at the end of this chapter.

You may want to use the flowchart to assist you in selecting the proper statistical test. To use the flowchart, first determine the *scale* of measurement of the data in that scenario. Next, decide whether the research question or hypothesis in that scenario predicts a *difference* or *association*. Next, decide whether there is one or more groups or variables in the study and whether they are *independent* or *paired*. The answers to all of these questions should help you choose the right statistical test.

In deciding which statistical test to use to answer the research questions and to analyze the data in the scenarios that follow, choose from these statistical tests:

1. Pearson correlation
2. A t test for *independent samples*
3. A t test for *paired samples*
4. A t test for a *single sample*
5. Chi square *test of best fit*
6. Chi square *test of independence*
7. One-way ANOVA
8. Two-way ANOVA
9. Simple regression
10. Multiple regression

In order to help you get the "hang of it," start by reading the two examples that follow. The examples include answers to the questions about the scale of the data, stated or implied hypothesis in the study, the groups used in the study, and some hypothetical data. As you work through the 14 passages, we suggest that you ask yourself the same questions as those in the two practice exercises and create some hypothetical data points.

Examples

Example 1

Many studies comparing on-the-job sales training techniques to in-class sales training techniques have found that companies using on-the-job techniques have salespeople who are more productive. A new sales trainer believes her in-class technique will be more successful than on-the-job training. She decides to test the two techniques by using the on-the-job training technique with half (30) of her new trainees in the next training class and her new in-class technique with the other half (30) of the trainees in the new class. After the training occurs, the sales manger tracks the number of new clients each of the trainees acquires for the company during the first year of selling. The trainer wants to compare the number of new clients each group of new salespeople acquired using the on-the job or in-class techniques.

Answer:

1. **Scale**: *Ratio*. The measure used in the study is the number of new clients each salesperson acquired for the company. It is possible a salesperson could have landed no (zero) new clients.

2. **Hypothesis**: A hypothesis of *difference*. The trainer predicts that the salespeople using her new in-class training technique would acquire a significantly higher number of new clients than those using the on-the-job training technique.

3. **Groups**: There are two groups of salespeople who are independent of each other.

4. What might the data look like?

On-The-Job Training	In-Class Training
45	34
50	29
38	41
Mean = 44.33	Mean = 34.66

> **Solution**: Since the means of two independent groups are being compared, using ratio scale data, use the *t* test for *independent samples*.

Example 2

In many states in the United States, a portion of the residents' property tax is used to pay for public education. School board members all over the country have noticed that district residents who have school-age children attending the district's schools are more likely than other residents to support a tax increase to improve education. In a suburban school district, a non-binding referendum about raising taxes to pay for education in the district is put on the ballot. The voters are asked to indicate their support or opposition to the referendum by marking the ballot with a *Yes* or *No* vote. The voters are also asked to indicate whether they have school-age children in the district's schools. The responses of the voters with and without school-age children are compared to determine whether those with children in the district's schools are more likely to support the tax increase, compared with voters who do not have children in the district's schools.

Answer:

1. **Scale**: *Nominal*. The study compares two groups of voters and their responses. The response choices are *Yes* and *No*.

2. **Hypothesis**: A hypothesis of *difference*. The hypothesis predicts that residents *with* school-age children would be more supportive of the referendum compared with residents *without* school-age children in the district.

3. **Groups**: There are two groups that are independent of each other. One group is voters with school-age children in the district's schools and the other group is voters without school-age children in the district's schools.

4. What might the data look like?

Group	Increase Taxes	
	Yes	No
Have children in the district	87	13
Don't have children in the district	21	79

> **Solution**: Data are presented in a form of *frequencies*; therefore, use the *chi square test*. Since there are two variable (groups and response choices), use the *chi square test of independence*.

Scenarios

1. An accountant wants to test whether his clients' accounts receivable levels are related to their profitability. The accountant selects a random sample of 200 client companies and compares their levels of accounts receivables in dollars with their profits (or losses) in dollars for the same time periods.

2. Past economics research has shown that males and females have different attitudes towards future economic conditions. Women generally have been shown to be more optimistic than men. To assess whether this difference still exists, an economist administered a consumer confidence survey to a sample of 100 men and 100 women to compare their attitudes towards the future of the economy. The consumer confidence survey is scored on a scale of 20-100 points.

3. The director of a college's food services is considering the addition of new items to the cafeteria menu. One of the new items is a green salad topped with strips of grilled chicken breast. After tasting the salad, the college's students, faculty, and administrators who eat at the cafeteria are asked to indicate their preference by circling one of the following options: (a) *Add it to the menu*, (b) *Do not add it to the menu*, and (c) *No opinion*. The director of food services analyzes the data to determine if there are differences in the numbers of students, faculty, and administrators who chose each of the three response options.

4. A statistics instructor at a business college has noticed that accounting and finance students seem to have more positive attitudes toward statistics compared with management and marketing students. The professor administers the *Statistics Attitudes Inventory (SAI)* scale to all students on the first day of the fall semester. The inventory contains 20 Likert-scale items with responses ranging from *Strongly Disagree* to *Strongly Agree*. The responses of accounting, finance, management, and marketing students are then compared to determine if there are significant differences in attitudes between the four groups of students.

5. A manager of a large manufacturing facility decides to implement a new quality management program. To assess whether the new quality management program is effective, the manager is going to test the quality of the products manufactured before and after the quality program is implemented. The quality test measures the percentage of finished units that have zero defects. The manager hypothesizes that there will be a significant increase in the *quality* of the products on the posttest compared with the pretest scores on the *quality* test.

6. A finance manager wants to know how good a predictor the Consumer Price Index (CPI) is at predicting a change in his company's stock price. He hypothesizes that a change in the CPI will result in a comparable change in his company's stock approximately one month later. To determine this, the manager tracks the percentage changes in the CPI and then compares it with his company's percentage change in stock price one month later. He is hoping the CPI can serve as a good predictor of future changes in his company's stock price.

7. The engineers in a large multinational corporation claim that the salespeople get paid more than employees in other departments. The CEO assures the employees that there is no significant difference in annual salaries between employees from different departments. The CEO conducts a study to compare the mean annual salaries of the 32 salespeople to the *mean* of the annual salaries of all the 530 employees of the company.

8. Many people view television as a major contributor to the shopping habits of teenagers. A marketing researcher decides to investigate whether there is a relationship between the number of hours children watch television and the amount of money they spend shopping. For two weeks, teenagers record the number of hours they watch television and how much they spend shopping. The researcher can now analyze the data (TV viewing hours and money spent shopping) and determine whether there is a relationship between these two variables.

9. Students in a statistics class are learning about probability. They conduct an experiment with a four-sided spinner. The students hypothesize that the spinner would land an equal number of times on each of the four sides. To test their hypothesis, the students spin the spinner 200 times and record the outcomes. They then compare their observed results to those that are expected. Since there are four sides, the students expect the spinner to land 50 times on each side.

10. A product manager is trying to determine whether to use a higher or lower price on a new product. She decides to test market the higher price in one city and the test market the lower price in another city. The two cities chosen to test the prices are similar in terms of their demographics and other factors. For four weeks the manger kept track of the sales volume in each of the two test cities. The manager compares the mean sales volume for each test city to determine which price is more successful.

11. Research to date has documented that there is a gender gap in computer use, in computer classes, and in employment in the field of Information Technology (IT). Some say that this disparity can be attributed in part to the fact that most electronic games are oriented toward boys' interests. A study is conducted with a randomly selected group of 250 teenage boys and 250 teenage girls who are given two electronic games to play. Two electronic games are tested: Game A is an adventure game that requires competition among the players, and Game B is an adventure game that requires collaboration between the players. Half of the boys and half of the girls are given Game A and the other half is given Game B. The researchers want to find out which computer game seems to appeal more to teenagers and whether there are gender differences in preferences and attitudes toward the two games. A 20-item survey measuring attitudes and opinions is administered to the teenagers after they play with their assigned game. Responses to each item on the questionnaire include four choices, ranging from *I liked it a lot* (4 points) to *I did not like it at all* (1 point). Scores on the survey items are added to create a total attitude score.

12. The directors of admissions in a large MBA program want to re-examine four variables currently used to select students for admission into the MBA program. They want to determine whether these variables are good predictors of students' success in the program, as measured by the students' graduate school GPA. The variables that are used for selecting students are: (a) the Verbal score on the Graduate Management Admission Test (GMAT), (b) the Quantitative score on the GMAT, (c) the Analytical Writing score on the GMAT and (d) the students' undergraduate GPA. The records of 500 randomly selected students who have completed their graduate studies are used to test how well the four predictors predicted the students' graduate school GPA.

13. A manager claims that employees who have worked at his company for a longer period of time tend to be more satisfied with their jobs compared with newer employees. Other managers at the company disagree with this opinion and predict that there will not be significant differences in the attitudes of employees, regardless of how long they have worked for the company. To confirm their opinion, they examine results from a survey that is administered annually to employees who have worked 1, 5 and 10 years for the company. The survey is designed to measure the level of employees' satisfaction with their jobs. The survey includes 30 questions with responses to each question ranging from 1 (*Very Dissatisfied*) to 4 (*Very Satisfied*). A mean score of overall satisfaction is obtained for each of the four groups of employees.

14. An accountant has noticed that her large clients appear to have different types of accounting problems than her smaller clients. To test whether there is a difference in the *type* of accounting problem based on the size of the client, 75 of her clients were asked to list their major problems. The accountant then classifies the clients' problems into five categories: *accounts payable*, *accounts receivable*, *cash flow*, *taxes*, and *debt*. The accountant compares the clients' problems viewed by smaller and larger companies to determine if differences exist between the two groups of clients.

Answers

Scenario 1
The association between two measures obtained for the same group of clients is assessed; therefore, *correlation* should be used. And, because the scale of measures is ratio, use the **Pearson correlation**.

Scenario 2
The mean scores of men and women taking the consumer confidence survey are being compared; therefore, use the *t* **test for independent samples**.

Scenario 3
Data are presented in the form of frequencies; therefore, use the *chi square test*. Since are there are two variables (type of cafeteria customer and response choices), use the **chi square test of independence**.

Scenario 4
The means of four independent groups (four college majors) are being compared; therefore, use the **one-way ANOVA**.

Scenario 5
Two means that are obtained for output from the same manufacturing plant are being compared; therefore, use the *t* **test for paired samples**.

Scenario 6
The percentage change in the CPI is used to predict the manager's future percentage change in the company's stock prices; therefore, use **simple regression**.

Scenario 7
A mean of one group (salespeople) is being compared to the mean of the population (all other employees); therefore, use the **single sample *t* test**.

Scenario 8
The association between two measures (number of hours children watch TV and their shopping spending) obtained for the same group of people is assessed; therefore, use the **Pearson correlation**.

Scenario 9
Data are presented in a form of frequencies; therefore, use the *chi square test*. Since only one variable is used (the sides of a spinner) and the expected frequencies are of equal probability, use the *goodness of fit* **chi square with *equal* expected frequencies**.

Scenario 10
The means of two independent groups (two prices in test cities) are being compared; therefore, use the ***t* test for independent samples**.

Scenario 11
There are two independent variables (*gender* and *type of game*), each with two levels. Therefore, there are four independent groups in the study. The means of the four groups are compared; therefore, use the **two-way ANOVA**.

Scenario 12
All five measures are obtained for the same group of students. Four of these measures (GMAT–Verbal, GMAT–Quantitative, GMAT—Writing Analysis, and undergraduate GPA) are used to predict the fifth measure (graduate GPA); therefore, use **multiple regression**.

Scenario 13
The means of three independent groups (those employed 1, 5 and 10 years) are being compared; therefore, use the **one-way ANOVA**.

Scenario 14
The data are presented in a form of frequencies; therefore, use the *chi square test*. Since there are two variables (*size of client* and the *type of accounting problem*), use the **chi square test of independence**.

List of Statistical Symbols

H_A	Alternative (research) hypothesis; also represented by H_1
H_0	Null hypothesis
p	Probability; level of significance
α	Probability level set at the beginning of the study
df	Degrees of freedom
ES	Effect size
$SE_{\bar{X}}$	Standard error of the mean
CI	Confidence interval
X	Raw score
N	Number of people in a group (or population)
n	Number of people in a group (or sample)
Σ	Sum of (the Greek letter sigma, upper case)
$\bar{X}$	Mean of sample
μ	Mean of population (the Greek letter mu)
S	Standard deviation (SD) of sample
S^2	Variance of sample
σ	Standard deviation (SD) of population (the Greek letter Sigma, lower case)
σ^2	Variance of population
z	z score
r	Pearson's correlation coefficient (also an index of reliability)

Y'	Predicted Y score (in regression)
b	Slope (or *coefficient*; in regression)
a	Intercept (or *constant*; in regression)
S_E	Standard error of estimate (in regression)
R	Multiple correlation coefficient
R^2	Coefficient of determination of multiple correlation (in regression)
t	*t* value
F	*F* ratio
K	Number of groups
SS	Sum of squares (in ANOVA)
MS	Mean squares (in ANOVA)
HSD	Tukey's honestly significant difference (in ANOVA)
χ^2	Chi square value
SEM	Standard error of measurement

Glossary

A-B-A single-case designs: Single-case experimental designs with three phases: **A** (baseline); **B** (intervention); and **A** (a second baseline, after the intervention is withdrawn). Multiple data points are used at each phase to obtain a stable measure of the target behavior. (Ch. 1)

Action research: Practitioner research; research that is undertaken to solve a problem by studying it, proposing solutions, implementing the solutions, and assessing the effectiveness of these solutions. The process of action research is cyclical; the researcher continues to identify a problem, propose a solution, implement the solution, and assess the outcomes. (Ch. 1)

Alternate forms reliability: An approach used to assess the degree of consistency between two forms of the same test. (Ch. 13)

Alternative hypothesis: A prediction about the expected outcomes of the study that guides the investigation and the design of the study. The alternative hypothesis is represented by H_A or H_1. Often, the alternative hypothesis is simply referred to as *the hypothesis*. It predicts that there would be some relationship between variables or a difference between groups or means. (Ch. 2)

Analysis of variance (ANOVA): A statistical test used to compare the means of two or more independent samples and to test whether the differences between the means are statistically significant. (Ch. 11)

Applied research: Research that is aimed at testing theories and applying them to specific situations. Based on previously developed theories, hypotheses are then developed and tested in studies classified as applied research. (Ch. 1)

Bar graph (bar diagram): A graph with a series of bars that do not touch that is used to display *discrete* and independent categories or groups. The bars are often ordered in some way (e.g., from the highest to the lowest). (Ch. 3)

Basic research: Research that is conducted mostly in labs, under tightly controlled conditions, and its main goal is to develop theories and generalities. This type of research is not aimed at solving immediate problems or at testing hypotheses. (Ch. 1)

Box plot (also called **box-and-whiskers**): A graph that is used to show the median and spread of a set of scores using a box and vertical lines. The two middle quartiles are located *within* the box and a horizontal line inside the box shows the location of the median. The two extreme quartiles are displayed using the vertical lines (the "whiskers") *outside* the box. (Ch. 3)

Causal comparative (ex post facto) research: Research designed to study cause-and-effect relationships, where the independent variable is not manipulated because it occurred prior to the start of the study, or it is a variable that cannot be manipulated. (Ch. 1)

Chi-square test: A nonparametric statistical test that is applied to categorical or nominal data where the units of measurement are *frequency* counts. *Observed* frequencies gathered in a study are compared to *expected* frequencies to test whether the differences between them are significant. The chi-square test statistic is represented by χ^2. There are two types of chi square tests: (a) The **goodness of fit** chi square test that is used with one independent variable; and (b) chi square **test of independence** that is used with two independent variables. (Ch. 12)

Class intervals: Equal-width groups of scores in a distribution. (Ch. 3)

Coefficient alpha: An approach to calculate the reliability of an instrument using scores from a single testing; also known as **Cronbach alpha**. (Ch. 13)

Coefficient of determination (r^2): An index used to describe the proportion of variance in one variable (usually the criterion) that can be explained by differences in the other variable (usually the predictor). (Ch. 8).

Concurrent validity: The correlation between scores from an instrument and scores from another well-established instrument that measures the same thing. (Ch. 14)

Confidence interval (CI): A range within which we would expect to find, with a certain level of confidence (e.g., 95%), the population value we want to estimate from our sample. The interval includes two boundaries: a lower limit (CI_L) and an upper limit (CI_U). (Ch. 2)

Constant: A measure that has only one value. (Ch. 2)

Construct validity: The extent to which a test measures and provides accurate information about a theoretical trait or characteristic. (Ch. 14)

Content validity: The degree to which an instrument measures behaviors and content domain about which inferences are to be made; the extent of the match between the test and the content it is intended to measure. (Ch. 14)

Continuous variable: A variable that can take on a wide range of values and contain an infinite number of small increments. (Ch. 2)

Convenience (or incidental) sample: A sample that is chosen for the study by the researcher because of its convenience. (Ch. 2)

Correlation: The relationship or association between two or more paired variables. (Ch. 8)

Correlation coefficient: An index indicating the degree of association or relationship between two variables. The coefficient can range from -1.00 (perfect negative) to +1.00 (perfect positive). The most commonly used coefficient is **Pearson r**. (Ch. 8)

Counterbalanced designs: Time-series experimental designs where different interventions are tested with intact groups by applying the same interventions to the groups in different order. (Ch. 1)

Criterion-referenced (CR) test: A test used to compare the performance of an individual to certain criteria. (Ch. 7)

Criterion-related validity: The degree to which an instrument is related to another measure, called the criterion. See *concurrent validity* and *predictive validity* for types of criterion-related validity. (Ch. 14)

Critical value: A value of a test statistic that is found in statistical tables of critical values, which are associated with different statistical tests. The computed test statistics are compared to the appropriate critical values in order to make decisions whether to retain or reject the null hypothesis. (Ch. 2)

Cross-sectional designs: Nonexperimental designs conducted to study how individuals change and develop over time by collecting data at one point in time on different-age individuals. (Ch. 1)

Cumulative frequency distribution: A distribution of scores that shows the number and percentage of scores *at* or *below* a given score. The distribution includes the following: scores, frequencies, percent frequencies, cumulative frequencies, and cumulative percent frequencies. (Ch. 3)

Degrees of freedom (*df*): In most cases, the degrees of freedom are *n*-1 (the number of people in the study, minus 1), although there are some modifications to this rule in some statistical tests. In most studies, degrees of freedom relate to the sample size. (Ch. 2)

Dependent variable: An outcome measure in an experimental study designed to measure the effectiveness of the intervention. (Ch. 1)

Descriptive research: Studies aimed at studying phenomena as they are naturally occurring, without any manipulation or intervention. (Ch. 1)

Descriptive statistics: Procedures used to classify, organize, and summarize numerical data about a particular group of observations. There is no attempt to generalize these statistics, which describe only one group, to other samples or populations. (Ch. 2)

Deviation score: The distance between each score in a distribution and the mean of that distribution, expressed as $X - \bar{X}$. (Ch. 5)

Differential selection: A threat to internal validity; refers to studies where pre-existing group differences may contribute to different performance on the dependent variable. (Ch. 1)

Directional hypothesis: A prediction that states the *direction* of the outcome of the study. In studies where group differences are investigated, a directional hypothesis predicts which group's mean would be higher. In studies that investigate relationships between variables, a directional hypothesis predicts whether the correlation will be positive or negative. (Ch. 2)

Discrete variable: A variable that contains a finite number of distinct values between any two given points. (Ch. 2)

Effect size (ES): An index that is used to express the strength or magnitude of difference between two means or the strength of association of two variables. The comparison of the means is done by converting the difference between the means into standard deviation units. Effect size can also be used to assess the strength of the association between two variables by using the correlation coefficient (r) or a square of the correlation coefficient (r^2, or R^2). (Ch. 2)

Experimental research: Research designed to study cause-and-effect relationships by manipulating the *independent* variable (i.e., the cause) and observing possible changes in the *dependent* variable (the effect, or outcome). Experimental research is designed to assess the effectiveness of a planned intervention on groups or individuals. (Ch. 1)

External validity: The extent to which the results of the study can be generalized and applied to other settings, populations, and groups. (Ch. 1)

Extraneous variable: A variable that presents a threat to the study's internal validity; an uncontrolled variable that can present a competing explanation to the impact of the planned intervention. (Ch. 1)

***F* value (or *F* ratio):** A test statistic used in the analysis of variance (ANOVA). It is computed by dividing two variance estimates by each other. (Ch. 11)

Face validity: The extent to which an instrument *appears* to measure what it is intended to measure. (Ch. 14)

Factorial ANOVA: A general name for ANOVA with two or more independent variables. (Ch. 11)

Frequency distribution: A distribution of scores that are ordered and tallied. (Ch. 3)

Frequency polygon: A graph that is used to display frequency distributions. The bell-shaped normal distribution is a special case of a frequency polygon with a large number of cases. (Ch. 3)

Grade equivalent (GE): A scale that is used to convert raw scores to grade level norms by expressing scores in terms of years and months. (Ch. 7)

Hawthorne Effect: A threat to external validity whereby the behavior of the study's participants may be affected by their knowledge that they participate in a study, rather than by the planned intervention. (Ch. 1)

Histogram: A graph that contains a series of consecutive vertical bars used to display frequency distributions. (Ch. 3)

History: A threat to internal validity; refers to events that happened while the study takes place that may affect the dependent variable. (Ch. 1)

Hypothesis: A prediction about the outcome of the study; an "educated guess." (Ch. 2)

Independent samples *t* test: A *t* test used to compare the mean scores of two groups that are independent of each other. (Ch. 10)

Independent variable: The intervention (or treatment) in experimental studies; a grouping variable in nonexperimental studies. (Ch. 1)

Inferential statistics: Procedures that involve selecting a sample from a defined population and studying that sample in order to draw conclusions and make inferences about the population. The sample that is selected is used to obtain sample *statistics* that are used to estimate the population *parameters*. May also be called *sampling statistics*. (Ch. 2)

Instrumentation: A threat to internal validity; refers to the level of reliability and validity of the instrument being used to assess the effectiveness of the intervention. (Ch. 1)

Interaction: A situation in factorial ANOVA where one or more levels of the independent variable have a different effect on the dependent variable when combined with another independent variable. (Ch. 11)

Internal consistency methods: Approaches used to assess the reliability of an instrument using scores from a single administration of the instrument. (Ch. 13)

Internal validity: The extent to which observed changes in the dependent variable (outcome measure) can be attributed to the independent variables (the intervention); the extent of control over the extraneous variables (Ch. 1)

Inter-rater reliability: A method to assess the degree of consistency and agreement between scores assigned by two or more raters or observers who judge or grade the same performance or behavior. (Ch. 13)

Interval scale: A measurement scale with observations that are ordered by magnitude or size with equal intervals between the different points. (Ch. 2)

John Henry Effect: A threat to external validity; refers to a condition where the intervention does not seem to be effective because control group members perceive themselves to be in competition with experimental group members and therefore perform above and beyond their usual level. (Ch. 1)

Level of significance (p level): The level of error associated with rejecting a null hypothesis; a probability that the study's results were obtained purely by chance. (Ch. 2)

Line graph: A graph used to show relationships between two variables through lines that connect the data points. The *horizontal* axis indicates values that are on a continuum and the *vertical* axis can be used for various types of data. (Ch. 3)

Longitudinal studies: Nonexperimental designs conducted to measure changes over time by following the same group of individuals. (Ch. 1)

Maturation: A threat to internal validity; refers to physical or mental changes experienced by the study's participants while the study takes place. (Ch. 1)

Mean: The most commonly used measure of central tendency that is obtained by adding up the scores and dividing the sum by the number of scores; also called the **arithmetic mean**. (Ch. 4)

Mean squares (MS): In ANOVA, there are different variance estimates. For example, MS_W is the estimate of the variances *within* groups; MS_B is the estimate of the variance of groups around the total mean. (Ch. 11)

Measure of central tendency: A summary score; a single score that represents a set of scores. (Ch. 4)

Measurement: A process of assigning numbers to observations according to certain rules. (Ch. 2)

Median: A measure of central tendency that is the distribution's midpoint, where 50% of the scores are above it, and 50% are below it. (Ch. 4)

Mode: A score that occurs with the greatest frequency; a measure of central tendency. (Ch. 4)

Multiple correlation (R): An index of the combined correlation of the predictor variables with the criterion variable. (Ch. 9)

Nominal scale: A measurement scale where numbers are used to label, classify, or categorize data. The various points on the scale are not ordered. (Ch. 2)

Nondirectional hypothesis: A hypothesis that predicts that there would be a difference or relationship but the direction of the difference or association is not specified. (Ch. 2)

Nonexperimental research: A research study where no planned intervention takes place. Nonexperimental research is divided into two types: *causal comparative* (also called *ex post facto*) and *descriptive*. (Ch. 1)

Nonparametric statistics: Statistics that are used with interval and ratio scale data that fail to meet the assumptions needed for parametric statistics, or with ordinal and nominal data. Nonparametric statistics are easier to compute and understand, compared with parametric statistics. (Ch. 2)

Normal curve: A graphic presentation of a theoretical model that is bell-shaped. Various characteristics in nature are normally distributed and each normal distribution has its own mean and standard deviation. (Ch. 6)

Norming group: A group used to develop test norms with demographic characteristics similar to those of the potential test takers. (Ch. 7)

Norm-referenced test: A test that includes norms designed to compare the performance of examinees taking the test to the performance of similar individuals in a norming group who took the same test and whose scores were used to generate national or local norms. (Ch. 7)

Null hypothesis: A hypothesis that predicts that there would be no relationship between variables or no difference between groups or means beyond that which may be attributed to chance alone; represented by H_0. In most cases, the null hypothesis (which may also be called the **statistical hypothesis**) is not formally stated, but it is always implied. (Ch. 2)

Ogive: A graph used to depict cumulative frequency distributions (also called the **"S" curve**). (Ch. 3)

One-tailed test: Used when the alternative hypothesis (i.e., the study's main research hypothesis) is directional to decide whether to reject the null hypothesis. (Ch. 10)

Ordinal scale: A measurement scale where the observations can be ordered based on their magnitude or size and the intervals among the different points on the scale are *not* assumed to be equal. (Ch. 2)

p **(probability) level:** The level of significance; indicates the *probability* that we are making an error in rejecting a true null hypothesis. A probability level of 5% is commonly used to decide whether to consider the results statistically significant. (Ch. 2)

Paired samples *t* test: A *t* test that is used to compare the mean scores of two sets of scores that are paired. (May also be called a *t* test for *dependent, matched*, or *correlated* samples). (Ch. 10)

Parameter: A measure that describes a characteristic or a value of an entire population. (Ch. 2)

Parametric statistics: Statistics that are applied to data from populations that meet the following assumptions: The variables being studied are measured on an interval or a ratio scale; subjects are randomly assigned to groups; the scores are normally distributed; and the variances of the groups being compared are similar. When these assumptions are met, researchers are likely to use parametric tests that are more efficient and powerful than their nonparametric counterparts. (Ch. 2)

Pearson Product Moment Correlation Coefficient: A correlation coefficient that is used to indicate the strength or degree of association between two sets of scores from two variables that are measured on an interval or ratio scale and have a linear relationship.

Percentile band: An estimated range where the true percentile rank of a student's percentile rank is expected to be, usually reported with 68% confidence level. (Ch. 7)

Percentile rank: An index that describes the relative position of a person by indicating the percentage of people at or below that score. (Ch. 7)

Pie graph (or **pie chart**): A graph that looks like a circle that is divided into "wedges", or "segments." Each wedge represents a category or subgroup within that distribution. (Ch. 3)

Population: An entire group of persons or elements that have at least one characteristic in common. (Ch. 2)

Post hoc comparison: In ANOVA, it is a process of *multiple comparison* done *after* the completion of the study where all possible pairs of means are compared. (Ch. 11)

Practitioner research: Action research; research that is undertaken by practitioners to solve a problem by studying it, proposing solutions, implementing the solutions, and assessing the effectiveness of these solutions. Educators conduct practitioner research to study their practice in order to reflect, describe, predict, and compare. (Ch. 1)

Predictive validity: The extent to which an instrument can predict some future performance. (Ch. 14)

Pre-experimental designs: Designs classified as pre-experimental do not have a tight control over extraneous variables and their internal validity cannot be assured. (Ch. 1)

Qualitative research: Research that seeks to understand social or educational phenomena. The researcher focuses on one or a few cases that are studied in depth using multiple data sources that are subjective in nature. Qualitative research is context-based, recognizing the uniqueness of each individual and setting. (Ch. 1)

Quantitative research: Research that is conducted to describe phenomena or to study cause-and-effect relationships by studying a small number of variables, and using numerical data. Researchers conducting quantitative research usually maintain objectivity and detach themselves from the study's environment. (Ch. 1)

Quasi-experimental designs: Experimental designs where intact groups are used and where the groups being compared are not assumed to be equivalent at the beginning of the study. (Ch. 1)

Range: A measure of spread (or variability) that indicates the distance between the highest and the lowest scores in the distribution. (Ch. 5)

Ratio scale: A measurement scale where the observations are ordered by magnitude, with equal intervals between the different points on the scale and an absolute zero. (Ch. 2)

Raw score: A score obtained by an individual on some measure that is not converted to another measure or scale. (Ch. 4)

Regression: A statistical technique used for estimating scores on one variable (the dependent variable, or criterion) from scores on one (or more) variable (the independent variable, or predictor). When one variable is used to predict another, the procedure is called *simple regression*, and when two or more variables are used as predictors, the procedure is called *multiple regression*. (Ch. 9)

Regression line: A line of best fit on a scattergram where the predicted scores are expected to be. (Ch. 9)

Reliability: The level of consistency of an instrument and the degree to which the same results are obtained when the instrument is used repeatedly with the same individuals or groups. (Ch. 13)

Research: A systematic inquiry that includes data collection and analysis. The goal of research is to describe, explain, or predict present or future phenomena. There are several ways to classify research and each approach looks at research from a different perspective. (Ch. 1)

Sample: A small group of observations selected from the total population for the purpose of making inferences about the population. A sample should be *representative* of the population, because information gained from the sample is used to estimate and predict the population characteristics that are of interest. (Ch. 2)

Sample bias: *Systematic*, rather than *random*, differences between the population and the selected sample; a systematic error in a sample. (Ch. 2)

Sampling error: A *chance* variation in the numerical values (e.g., mean) of a sample that occurs when we repeatedly select same-size samples from the same population and compare their numerical values. Sampling error is beyond the control of the researcher. (Ch. 2)

Scattergram: A graph used to depict the association (correlation) between two numerical variables. (Ch. 8)

Simple random sample: A sample where every member of the population has an equal and independent chance of being selected for inclusion. (Ch. 2)

Single-case (or **single-subject**) **designs:** Experimental designs where individuals are used as their own control. Their behavior or performance is assessed during two or more phases, alternating between phases *with* and *without* an intervention. (Ch. 1)

Single sample *t* test: A *t* test used to compare the mean of a sample ($\overline{X}$) to the mean of a population (μ). (Ch. 10)

Spearman-Brown prophecy formula: This formula is used to estimate the reliability of a test using the reliability computed for the half-length test. The Spearman-Brown prophecy formula is:

$$r_{full} = \frac{(2)\,(r_{half})}{1 + r_{half}}$$

Split-half method: A procedure for assessing the test reliability by dividing the items into two halves and correlating the scores from one half with the other. (Ch. 13)

Standard deviation (SD): A measure of spread in a distribution of scores. It is the mean of the distances of the scores around the distribution mean. The standard deviation is the squared root of the variance. The SD of the sample is S and the SD of the population is σ (the Greek letter Sigma, lower case). (Ch. 5)

Standard error of estimate: An index that estimates the amount of error expected in predicting a criterion score; the standard deviation of the differences between actual and predicted scores. (Ch. 9)

Standard error of measurement (SEM): An estimate of the error in a person's score on a test. (Ch. 13)

Standard error of the mean: The standard deviation of the sample means, expressed by the symbol $SE_{\bar{x}}$. (Ch. 2)

Standard score: A derived scale score that expresses the distance of the original score from the mean in standard deviation units. The most common standard score is the z score. (Ch. 6)

Stanine: A 9-point scale that is derived from the words "**standard nine**", with a mean of 5 and a standard deviation of 2. Stanines allow the conversion of percentile ranks into larger units. (Ch. 7)

Statistic: A measure that describes a characteristic of a sample. (Ch. 2)

Statistical regression: A threat to internal validity; refers to a phenomenon whereby people who obtain extreme scores on the pretest tend to score closer to the mean of their group upon subsequent testing, even when no intervention is involved. (Ch. 1)

Statistical significance: Most researchers use the convention whereby they report their findings as statistically significant if their computed probability level (p value) is 5% or less ($p \leq .05$). Reporting results as statistically significant means that the likelihood of obtaining these results purely by chance is low and that similar results would be obtained if the study was repeated. (Ch. 2)

Stratified sample: A sample that contains proportional representations of the population subgroups. To obtain a stratified sample, the population is first divided into subgroups (strata), then a random sample is selected from each subgroup. (Ch. 2)

Sum of squares (SS): In ANOVA, these are different sources of variability. The **within-groups sum of squares (SS$_W$)** is the variability *within* the groups. The **between-groups sum of squares (SS$_B$)** is the average variability of the means of the groups *around* the total mean. (May also be called **among-groups sum of squares**; abbreviated as SS$_A$.) The **total sum of squares (SS$_T$)** is the variability of *all* the scores around the total mean. (Ch. 11)

Systematic sample: A sample where every *Kth* member (e.g., every 5th person) is selected from a list of all population members. (Ch. 2)

t **test:** A statistical test used to compare two means. The means may be from two different samples, from paired samples, or from a sample and population. The scores used to compute the means should be measured on an interval or ratio scale and be derived from the same measure. (See independent samples *t* test, paired samples *t* test, and single sample *t* test.) (Ch. 10)

Testing: A threat to internal validity; refers to the potential effect that a pretest may have on the performance of people on the posttest. (Ch. 1)

Test-retest reliability: A procedure for assessing the reliability of a test by administering the test twice to the same group of examinees and correlating the two sets of test scores. (Ch. 13)

Time-series designs: Designs that are classified as quasi-experimental, where intact groups are tested repeatedly *before* and *after* the intervention. (Ch. 1)

True experimental designs: Experimental designs where the groups are considered equal because participants are randomly assigned to groups. (Ch. 1)

Tukey's honestly significant difference (HSD): The HSD value is used in the ANOVA procedure as part of the post hoc comparison to determine how large a difference between means should be to be considered statistically significant. (Ch. 11)

Two-tailed test: Used when the alternative hypothesis (i.e., the study's main research hypothesis) is nondirectional or stated as no difference between means or no association between variables. (Ch. 10)

Type I error: An error made by researchers when they decide to *reject* the null hypothesis (H_O) when in fact it is true and *should not be rejected.* (Ch. 2)

Type II error: An error made by researchers when they decide to *retain* the null hypothesis, when in fact it *should be rejected.* (Ch. 2)

Validity: The degree to which an instrument measures what it is supposed to measure and the appropriateness of specific inferences and interpretations made using the test scores. (Ch. 14)

Variable: A measured characteristic that can assume different values or levels. (Ch. 2)

Variance: A measure of spread in a distribution of scores, in squared units. It is the mean of the squared distances of the scores around the distribution mean. The variance can be obtained by squaring the standard deviation. The variance of the sample is S^2 and the SD of the population is σ^2 (the Greek letter Sigma, lower case, squared). (Ch. 5)

z score: A type of standard score that indicates how many standard deviation units a given score is *above* or *below* the mean for that group. The z scores create a scale with a mean of 0 and a standard deviation of 1. (Ch. 6)

Index

CPSIA information can be obtained at www.ICGtesting.com
Printed in the USA
LVOW09s1106090114

368749LV00001B/13/P

9 780761 838845